ANTHRACITE TRUST

AILEEN SALLOM FREEMAN

FOSI
Ltd

CREDITS: Cover Photograph - Richard Freeman
Editorial - Murray Freeman

NONFICTION

Chapters 1 through 8 appeared in Newsletter form from 1992 to 1994

FIRST EDITION
November 1994

Library of Congress Card Catalog Number: 94-61931

Published by FOSI, Ltd.
Paupack, Pennsylvania 18451-0184

Manufactured in the United States of America

ISBN 0-9644199-0-4

to Murray,
to Richard,
and to the achievements of the men
of whom I have written.

Foreword

The development of the anthracite coal industry in the United States began approximately 200 years after the first English Colonists arrived at Jamestown, Virginia and Plymouth, Massachusetts. Yet life in the coal region was as if two centuries of civilization had not intervened. Rough-hewn furniture and homespun cloth were the order of the day, rather than stately longcase clocks and brocades, although only 50 miles or so separated the two.

But the vigor of the men, women and children who moved into Pennsylvania's anthracite region was unmatched; in less than one generation they spanned the 200-year gap, and then went far beyond. It is this enormous energy and enthusiasm that this book discusses in an introductory way - introductory, because every name mentioned on these pages is worthy of a book in itself.

Not all aspects of the anthracite industry are reviewed, notably social conditions and analyses of individual businesses. Again, these are subjects worthy of deeper considerations.

The facts that company housing was provided for miners under lease-purchase agreements, that initial contracts specifically mandated support pillars in underground mining, and that legislation was passed protecting an animal population feared to be on the decline, are important points to remember when considering social issues of the age of the seemingly wanton industrial exploitation of man and environment - 19th century America. They are not, however, essential to the purpose of this book. Neither are the successes or failures of individual businesses, generally speaking. Again, other factors, particularly legal interpretations, are vital aspects of the business mergers, bankruptcies, or close-outs that sometimes occupied the attention of the men mentioned in this work. Again, except in one or two instances, complete and detailed business histories are not germane to the purpose of this book, nor is every activity these men participated in mentioned.

The aim of this book is to introduce, or perhaps for some readers to reintroduce, a group of very private men who structured their lives around America's rise to a unified nation and world power.

As for the format of this book, to make the reading of the thousands of documented facts as enjoyable as possible, there are no footnotes, and only a limited bibliography follows at the end. But because this is a reference book, and because the information has been gleaned mainly from primary and some secondary source materials, some quite lengthy, the publisher will retain copies of the basic documents as a special collection, and will provide specific bibliographic references upon written request.

Chapter 1

If you had been living here in northeastern Pennsylvania two hundred years ago, you would have been living in Connecticut!

Yes, that's right, you would have been a Connecticut property owner. That's because all of northern Pennsylvania belonged to Connecticut.

Starting on the west bank of the Delaware River, Connecticut land ownership stretched across the Endless Mountains into Ohio. Documents signed by English kings gave Connecticut land ownership to the Pacific Ocean!

Connecticut land became Pennsylvania land because of two English kings, one of whom had a very short memory. Charles I granted permission in 1620 for settlements in the Connecticut River Valley. His successor, Charles II, not only reaffirmed the Connecticut settlements but enlarged them. By royal charter issued in 1662, Charles II decreed Connecticut's boundary to be Narragansett Bay on the east, Long Island Sound on the south, and the Pacific Ocean on the west. The latitude for the northern boundary was an extension of the line separating Connecticut and Massachussetts.

Charles II, afterwards, apparently forgot what he had done for Connecticut. He granted, 19 years later in 1681, a royal patent to William Penn for land west of the Delaware River between Lord Baltimore's land grant on the south, and lands of New York on the north.

On paper, Penn and his aides had created maps establishing various land tracts as manors. One manor was named Wallenpaupack. It consisted of approximately 12,000 acres and had the Wallenpaupack Creek flowing through its center. *Today,* portions of northern Pike County and southern Wayne County make up Wallenpaupack Manor.

Closer to Philadelphia, Penn and his followers established baronies. The Welsh barony consisted of 40,000 acres. *Today,* the barony's original lands comprise almost all Montgomery County and a small part of Delaware County, suburbs of Philadelphia.

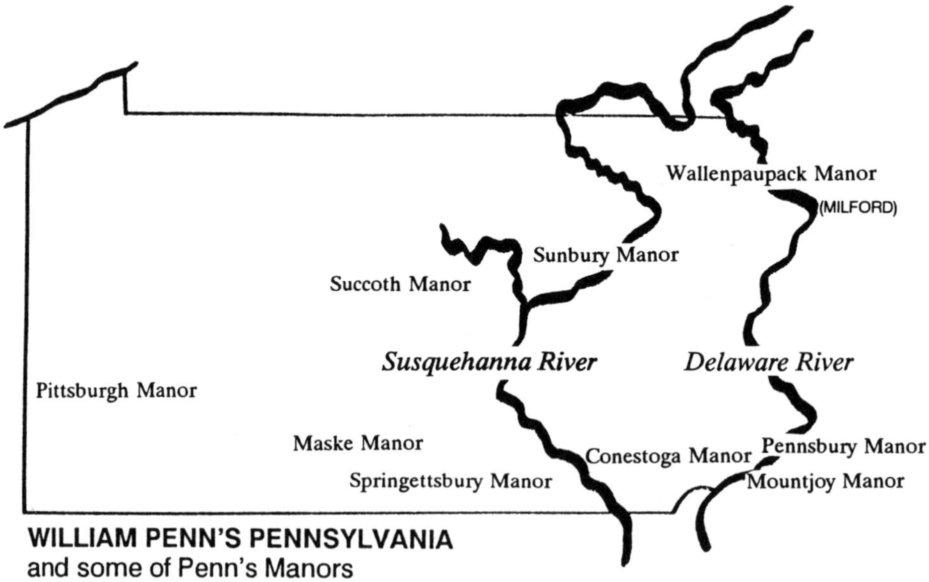

WILLIAM PENN'S PENNSYLVANIA
and some of Penn's Manors

Penn, a member of the religious Society of Friends, or Quakers, sold his manors, baronies, or portions of them, to fellow Quakers, or to anyone interested in settling in his "Holy Experiment" (a community with religious and political toleration). Many Philadelphians bought an interest in manor lands in addition to their urban holdings. However, the new urban owners preferred to remain in Philadelphia and very rarely visited their northern/western lands.

Settlement of Pennsylvania spread out from its southeast corner around Philadelphia. Except for explorers, fur traders, and missionaries, very few people identified as Pennsylvanians ever visited the northern one-third of Pennsylvania.

Connecticut, like Pennsylvania, had done nothing to settle its western wilderness. When settlers had first started to move into the Connecticut River Valley in the early 1600s, the land to the west, along the North (Hudson) River was known as New Netherland, a Dutch colony belonging to The Netherlands. In 1664 the Hudson River territory came until English control for the first time and renamed New York. No concession was given Connecticut for a strip of land through New York to connect the original

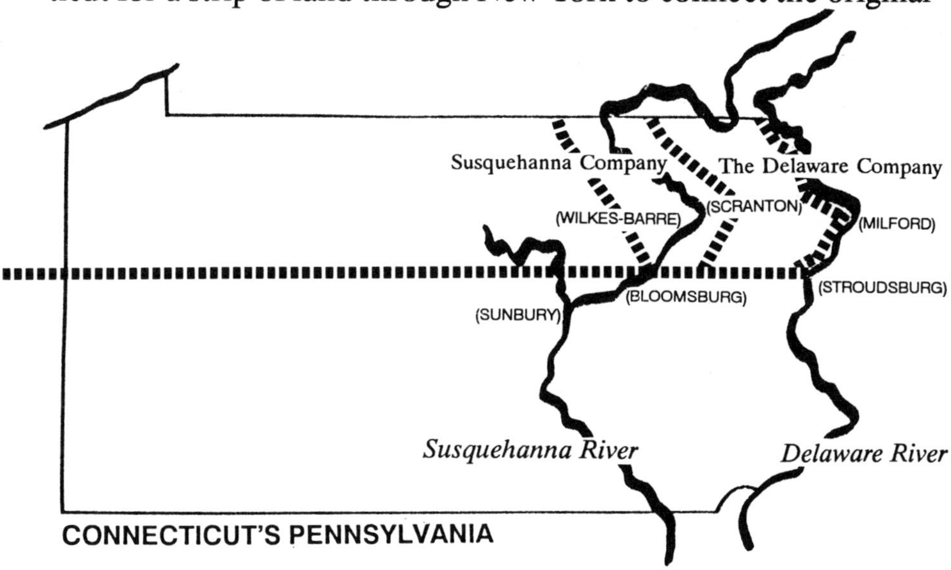

CONNECTICUT'S PENNSYLVANIA

Connecticut Colony with its western land. Except for explorers, fur traders and missionaries, very few people who were residents of Connecticut ever visited the western lands. Then in March 1753, specific plans for colonization were presented to the Connecticut Assembly by members of a newly formed land company, the Susquehanna Company.

One hundred persons, inhabitants of Windham County, presented their land company proposal to the Assembly. Shortly afterwards, another land company, The Delaware Company, was formed.

The Susquehanna Company was to develop a tract of land sixteen miles square on both sides of the Susquehanna River. The Delaware Company was to develop all the land west of the Delaware River to where it touched lands of the Susquehanna Company.

August of 1762 saw 150 people from Connecticut, members of the Susquehanna Company, moving towards the Susquehanna River to clear the land (*now* Wilkes-Barre area). They followed an old, well worn Indian path from the Delaware River.

How the Connecticut claimants and the Pennsylvanians had dealt, and would deal, with the native Americans determined the next series of events. European affairs and England's need for a formal relationship with the Indian people would also come into play.

Chapter 2

"A mighty thing, this our Great Peace. Have you, of across the water, had a greater vision?"

The Confederacy, the Iroquois Confederacy, was driven by a dream of universal peace. The Iroquois asked the white settlers,

"A mighty thing, this our Great Peace. Have you, of across the water, had a greater vision?"

Homelands of the Indian Nations comprising the Iroquois Confederacy were spread over most of Central New York. Five Indian Nations, the Mohawks, Onondagas, Senecas, Oneidas, Cayugas, and later, Tuscaroras, to make six nations, were (and still are) the Iroquois Confederacy.

Along the southern edge of Iroquois homelands, just over the Pennsylvania border in New York State, lay the Forbidden Path. The Seneca Nation patrolled the highway, safeguarding the rear approach to Iroquois Confederacy lands. No white man used the path.

From the path south, through the Endless Mountains of Pennsylvania, to the broad plains surrounding Harrisburg and Lancaster County, reaching to the Pennsylvania-Maryland border, the Iroquois Confederacy held suzerainty.

Originally, the Susquehannock Indians populated the area. In a power play, between Maryland and New York, for control of the Delaware River Bay, the Iroquois came to a conference. Held on the future site of Philadelphia, in March of 1677, the Iroquois induced the Susquehannocks to come and live among them.

The land where the Iroquois reigned but did not live is best described today as the Susquehanna River Watershed. With the Susquehannocks' migration north, the Susquehanna River's watershed land became a vast empty tract. The Iroquois worried that unwanted visitors, enemies, would move in. Their Confederacy embarked on a policy of inviting Indian tribes to settle there.

The Lenni Lenape Indians, known as the Delaware, along with their cousins, the Munsee, had claimed both sides of the Delaware River for their homelands. Delawares were found across New Jersey, as far east as the Atlantic Ocean. The Delaware (including the Munsee) had always lived in New Jersey/Pennsylvania. The Delawares would eventually be joined by a small band of Shawnee who had moved from Ohio, through Maryland, to settle above the Delaware Water Gap in Pennsylvania.

When the Delaware and Shawnee lost their original lands to the white settlers, the Iroquois were quick to offer the great Wyoming Valley (Scranton and Wilkes-Barre) and the Forks of the Susquehanna River/Shamokin (present day Sunbury) to the displaced refugees.

As can be seen in the following map, the Delaware River Basin area was not included in the lands over which the Susquehannocks roamed.

William Penn, at the time of the Iroquois' invitation to fellow tribes, was granted land in the new world in 1681. Drawing up plans for his colony, Penn ordered that not only were land titles to come from the King of England, but also from the American Indian in order to be valid. Every acre of Penn's woods would have to be purchased from its Indian owner.

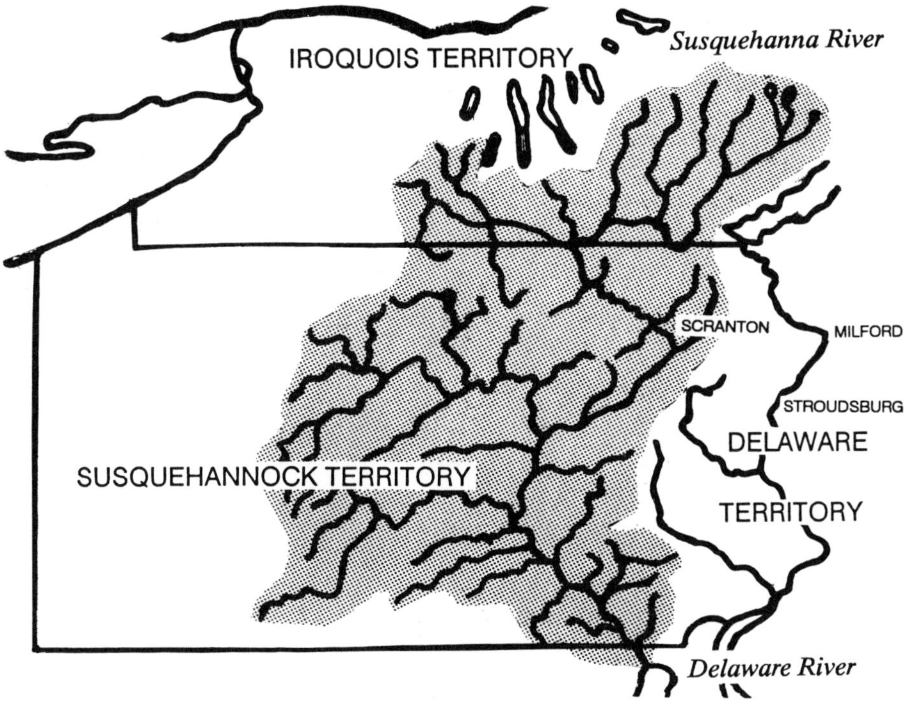

Thus, the Quaker who founded "The Peaceable Kingdom" (Penn) came to talk and deal with those who lived under "The Tree of Peace" (Iroquois).

Land purchases made by Penn and his agents, starting with the Philadelphia area in 1682, were successful models of cooperation and agreement between the "white man" and the "red man."

The Walking Purchase of 1737, however, changed everything.

Pennsylvanians, at the time, were pushing the boundary of "their country" further north and west with every purchase of land from the Indians. In March of 1737, Penn Family agents began proceedings for a "confirmation deed."

When buying Indian lands, the white man or Pennsylvanian purchasers would first obtain an "agreement deed." Usually the deed descriptions were vague, using phrases such as west "so far as a horse can travel in two summer days."

Afterwards, by the confirmation deed, the land sold would record the surveyor's precise distances.

The 1737 confirmation deed was for a purchase 51 years earlier, in 1686, for all the land between the Delaware River and Neshaminy Creek, northwest "extending as far as a man can go in one day and a half."

The Pennsylvanians had their eye on the Forks of the Delaware River, present day Allentown, Bethlehem, and Easton area. Their calculations showed that starting from Wrightstown (above today's Levittown, northeast of Philadelphia) by the banks of Neshaminy, a day and a half fast travel could bring the man to Jim Thorpe.

A right angle line (per the deed agreement) from the point where his run ended (Jim Thorpe) east to the Delaware River would intersect the Delaware near Lackawaxen (in Pike county).

More importantly, all the land around the Forks of the Delaware would then be included in the 1686 agreement deed.Three men began the one and one-half day journey, *running,* not walking. Edward Marshall, one of the three, went the farthest distance, 65 miles, to what is now called Jim Thorpe.The Indian observers for the confirmation were furious! For with the "Walk," the Delawares lost the last of their lands!

The Iroquois, older brother/uncle to the Delawares, offered the great Wyoming Valley for their new home.

Teedyuscung, recognized as chief by the Delawares, Munsees, Shawnees and Mahicans (this last from Hudson River area), accepted the Confederation's offer and moved his people to what is now Wilkes-Barre/Scranton area.

Later, at the height of the French and Indian War (1754-63), *Teedyuscung said the "Walk" was "not the principal cause that made us strike our Brethren, the English, yet it has caused the stroke to come harder than it otherwise would have come."*

Edward Marshall, for his part, returned to the area encompassed in his running, settling for a while at what is known today as Marshalls Creek, where Pennsylvania Routes 209 and 402 meet.

And for a while, Pennsylvania had a new, slightly expanded, border line.

The next time Pennsylvania's border was to change was after the Connecticut Yankees visited those who lived under *The Tree of Peace,* described as "*a great white pine tree 'rising to meet the sun'.*"

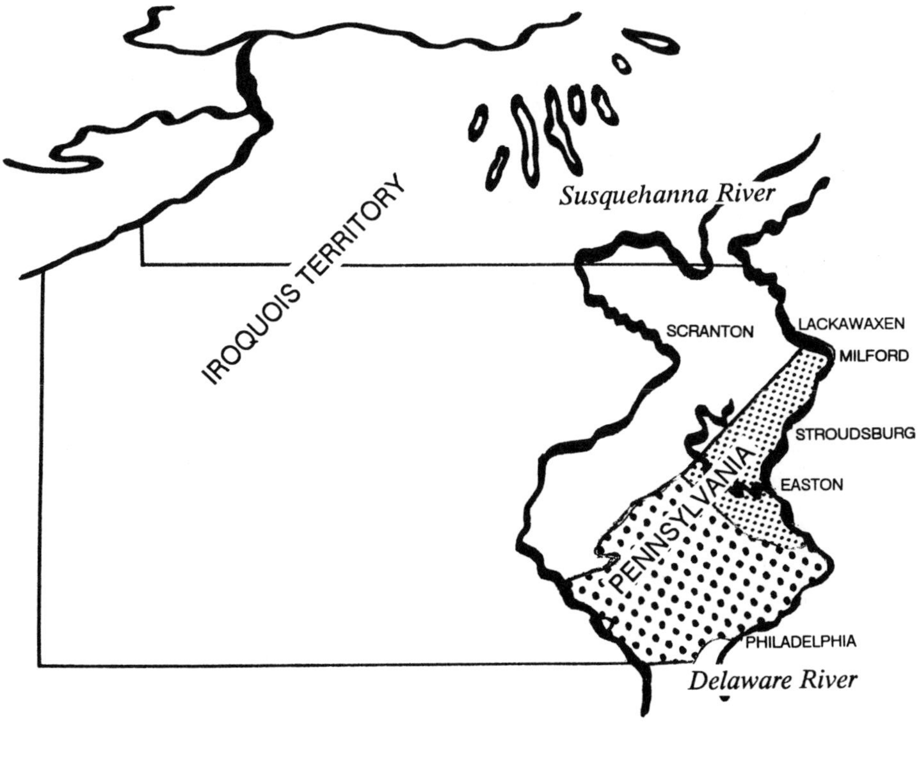

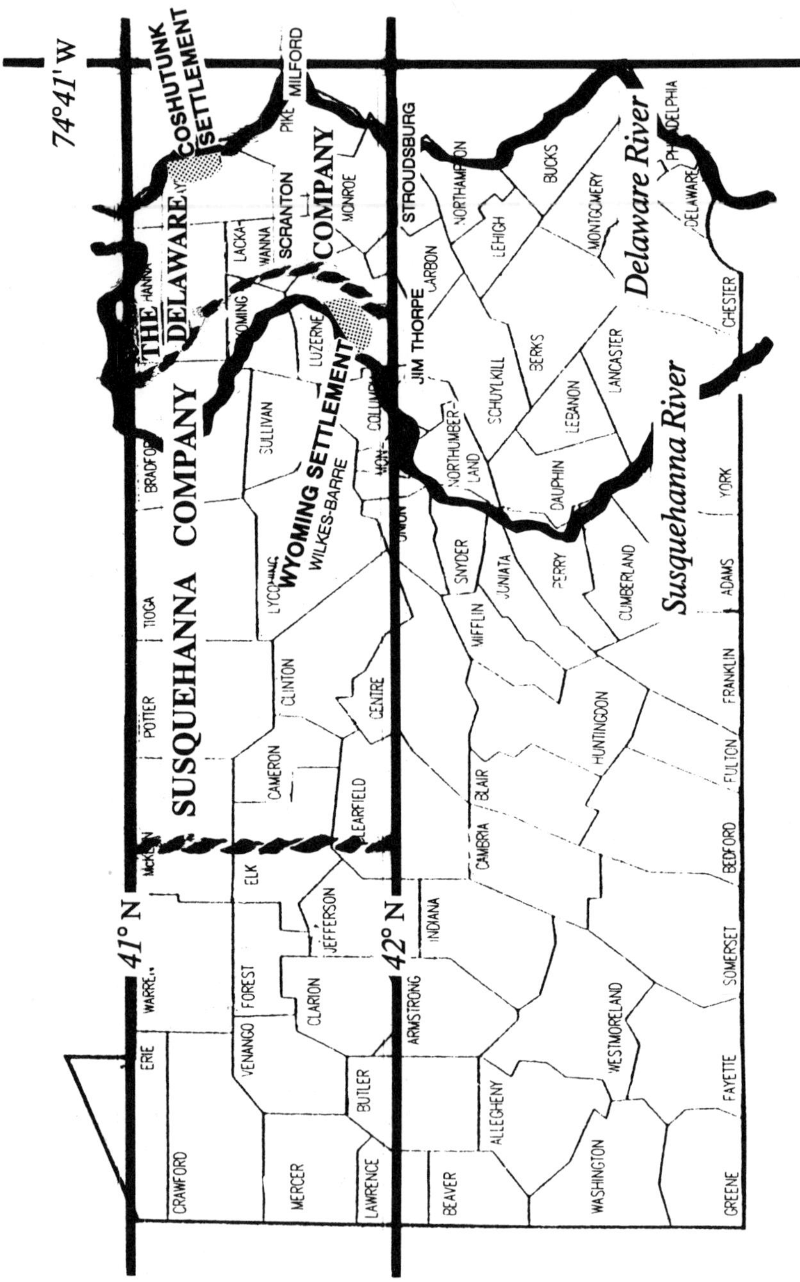

FIRST CONNECTICUT SETTLEMENTS

Chapter 3

Connecticut's title to land between 41° and 42° north latitude and west of the 74°41' meridian of west longitude was perfected.

He was probably the most powerful person in the American Colonies before the Revolution War of 1776. His name was Sir William Johnson, and he was powerful because he had the ear of king (George II) and government (Parliament) in England and the friendship of the American Indians.

It was Johnson who allied the Iroquois Confederacy with the British during the French and Indian War (1754-1763). This gave England victory over France in the New World while achieving the same supremacy in the concurrent Seven Years War in Europe.

Sir William, sometimes called Warraghiyagey, carved a baronial estate for himself out of the wilderness, approximately forty miles northwest of Albany, New York. The grounds of his manor house were always open to the Indians, and it was not unusual to see small encampments surrounding his home.

After the death of his German wife, Catherine, Johnson lived with Caroline, an Indian. He had children by both women and later by his housekeeper, Molly Brant.

In his official capacity of Superintendent of Indian Affairs in the North, colonists and Indians sought his participation and approval to every joint action.

So in 1754 it was to Albany that Connecticut men, representatives of the Susquehanna Company and The Delaware Company, traveled when Sir William convened a formal council meeting.

While the Congress was in session, the Connecticut men met with 18 Chief Sachems of the Iroquois Confederacy and offered them £2,000 New York currency ($5,000) for the northeastern corner of Pennsylvania. The purchase money was accepted and a deed was signed on July 11, 1754.

With this deed, the Connecticut men fulfilled the first two of three requirements to owning land in the American Colonies. The third requirement was "to occupy the land." (The first was a deed from the King of England, the second, a deed from the Indian owner.)

Immediately after the purchase, Pennsylvanians began a campaign to have the deed cancelled. On November 15, 1754, Pennsylvania [Lieutenant-] Governor Robert H. Morris wrote to Sir William Johnson and asked him to "lend the whole weight of his influence to render nugatory" the deed.

By January 15, 1755, the Indian Peter Hendrick, sent by Johnson, was speaking before Pennsylvania's Council of Government. "We will give you our assistance; but you know that we cannot destroy the deed ourselves...It must be the act of the Council of the Six Nations [Iroquois Confederacy]," he said.

For a while, time appeared to be on the side of Pennsylvania. Because of the French and Indian War, it was not until 1757 that a small band of Connecticut settlers, under their purchase rights in The Delaware Company, moved to settle on the west bank of the Delaware River at Coshutunk.

Coshutunk Valley is the area along the Delaware River between Callicoon, New York and the point at which the Lackawaxen River empties into the Delaware. It is partly in Pike County and partly in Wayne County, on the Pennsylvania side. The approximate location today of the center of the Connecticut/Coshutunk settlement is Damascus, in Damascus Township, Wayne County, Pennsylvania.

The following year, in October 1758, Sir William Johnson was urged to call another Congress. This he did, at Easton, resulting in "one of the most imposing assemblages ever beheld in Pennsylvania."

Teedyuscung, king of the Delawares, who came to Easton from his home at Wyoming (Wilkes-Barre), took an active part in the Congress, speaking eloquently, and "the different tribes at length became reconciled to each other."

The Congress "having continued 18 days, and all causes of misunderstanding between the English and Indians being removed, a general peace was concluded on the 26th day of October [1758]."

In the peace that followed, the Connecticut settlement at Coshutunk grew; and by 1760, there were 30 dwelling houses, a block-house for defense, a grist-mill, and a saw-mill at the site.

Two years later in 1762, with the French and Indian War almost at an end, a considerable number of men from Connecticut passed south of Coshutunk on their way to the great Wyoming Valley (Wilkes-Barre and Scranton). They sowed grain on the "flats" of the Susquehanna River, below Teedyuscung's village at Wilkes-Barre. Afterwards, the men returned to Connecticut for the winter. In the spring, they would be back with their families.

Occupation, fulfillment of the last of the three requirements for land ownership, had begun.

Connecticut's title to land between 41° and 42° north latitude and west of the 74°41' meridian of west longititude was perfected.

If the title was ever to be destroyed, "It must be the act of the Council of the Six Nations."

Chapter 4

"The Proprietaries [of Pennsylvania] obtained a deed from the Six Nations, of the Susquehanna lands, being the same previously claimed to have been conveyed to the Susquehanna Company [of Connecticut]."

It was the night of April 19, 1763. On the banks of the Susquehanna River, at a place called Wyoming (Wilkes-Barre), a hand reached out. The faint glow of the unextinguished brand it held gently touched a bark covered building. Not long after, flames of fire crackled skyward, the funeral pyre of Teedyuscung, chief of the Delaware Indians in Pennsylvania.

The Delaware chieftain and "twenty of the surrounding dwellings burst, almost at the same moment, into flames," all consumed by fire.

To this day, no one has determined who torched the home of Teedyuscung where he lay sleeping.

The Connecticut settlers were blamed.

The Connecticut settlers were blamed because they were close by, also on the banks of the Susquehanna River at Wilkes-Barre.

In the spring of 1763, before April 19, the Connecticut men who had sown grain the previous fall returned to Wilkes-Barre with their families. They brought household goods with them and prepared to build a permanent settlement.

They were unaffected by Teedyuscung's death, spending a quiet summer farming. At summer's end, they harvested an "immense" corn crop and were looking forward to "prosperity and happiness."

But on October 15, 1763, the new life came to an end. An Indian war party attacked, killing about twenty Connecticut men and carrying others off as captives. It was retribution for the death of Teedyuscung. Survivors of the attack fled eastward, eventually back to Connecticut.

The other Connecticut settlement, at Coshutunk on the Delaware River, was left untouched. Only the white man's presence at Wyoming (Wilkes-Barre) was eradicated.

From the very first encroachments of the white man into Indian territory, Iroquois and Delawares had spoken of Wyoming as their homeland. It was never to be given up. Land north of Blue Ridge Mountain was theirs.

(Delaware Water Gap, at Stroudsburg; Wind Gap, approximately seven miles south of Pennsylvania Route 209 on Route 33; Lehigh River; and the Susquehanna River are main openings in the Blue Mountain ridge, which runs southwest to northeast from Chambersburg, near the Pennyslvania-Maryland border, into New Jersey as the Kittatinny Mountains north of Newton.)

The earlier loss of land by the Walking Purchase of 1737 to Pennsylvania, and the sale of northeastern Pennsylvania to the Susquehanna Company and The Delaware Company of Connecticut, became political issues, used by the Iroquois Confederacy and Delaware Indians in their relationship with each other and in their dealings with the white man.

But regardless of what the Indians said, regardless of sales or treaties, it was soon apparent that neither the Delaware nor members of the Iroquois Confederacy had ever intended the white settler to intrude beyond the Blue Ridge.

On November 24, 1755, Gnadenhuetten had been torched.

At first, the incident was attributed to Indian unrest because of the ongoing French and Indian War.

Gnadenhuetten was located on the Lehigh River, north of the Blue Ridge near present day Jim Thorpe in Carbon County. It was the northernmost settlement by Pennsylvanians as they expanded outward from Philadelphia. It had begun nine years earlier as a Christian mission for Indian converts.

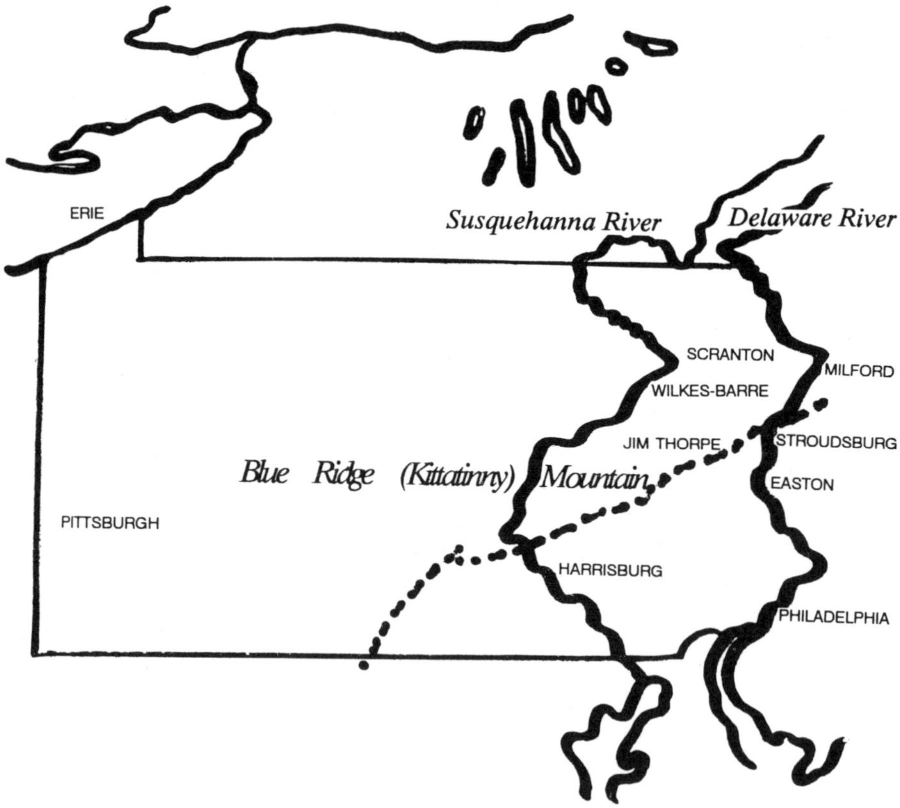

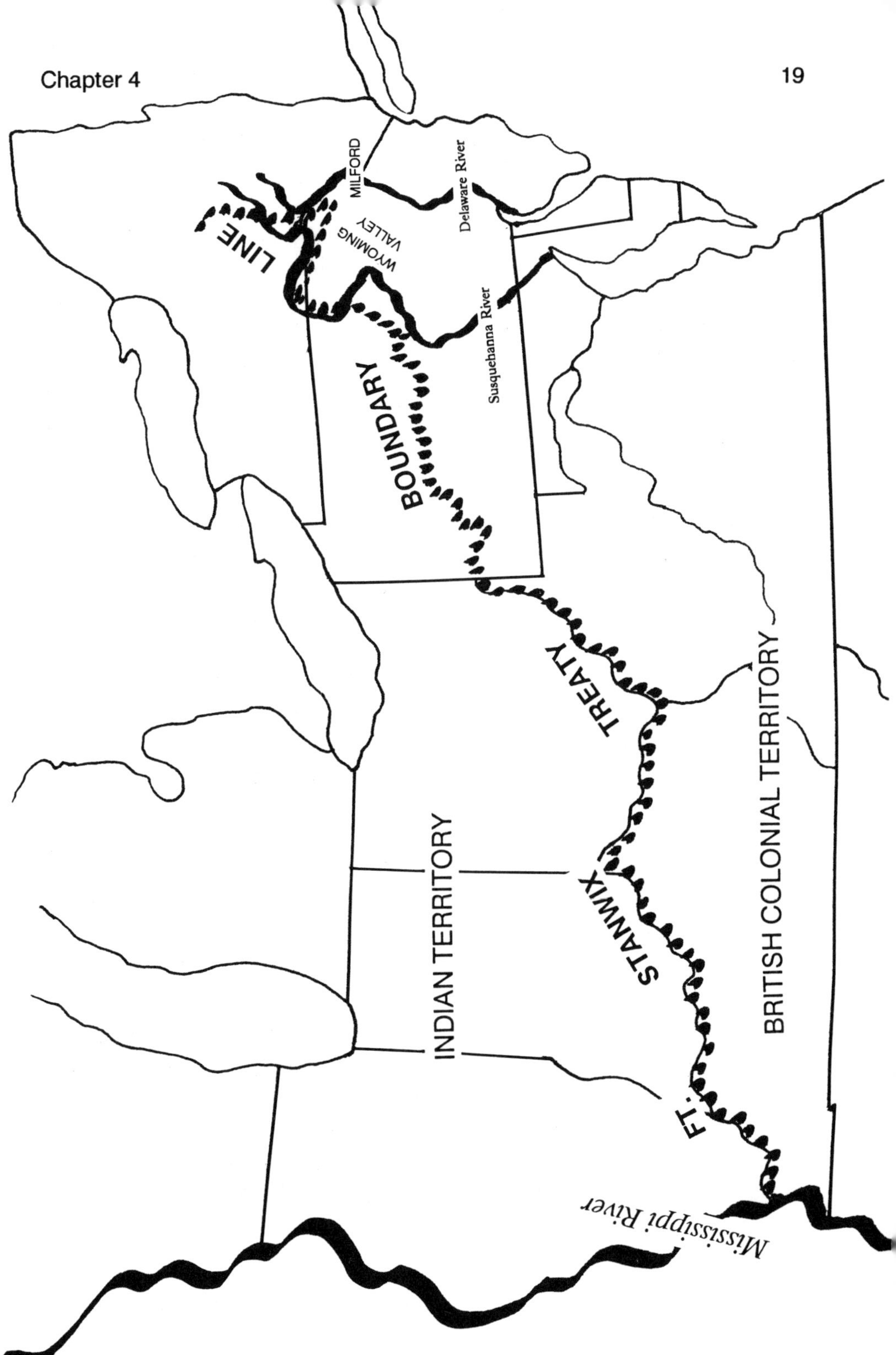
MILFORD
Delaware River
WYOMING VALLEY
Susquehanna River
FT. STANWIX TREATY BOUNDARY LINE
INDIAN TERRITORY
BRITISH COLONIAL TERRITORY
Mississippi River

The mission had been started by members of the Moravian Church, a Christian denomination from Germany that had founded Bethlehem, Pennsylvania.

One Moravian, G. M. Weiss, had also established an iron manufactory, Maria Forge, near Gnadenhuetten.

With the destruction first of Gnadenhuetten and then of the Susquehanna Company's settlement at Wilkes-Barre, Indians were once again the sole possessors of northeastern Pennsylvania.

A lone family, the Carters, also from Connecticut, had settled at today's Paupack, in Pike County, near the present day Paupack Methodist Church. During the French and Indian War, the Carters were killed and their house burned.

Coshutunk settlers, while harassed by Indian raids, as were all white settlers along the Delaware River, were not driven off the land as had happened at Gnadenhuetten, Wilkes-Barre, Paupack, and isolated single-family settlements throughout northeastern Pennsylvania.

When the French and Indian War was concluded by The Treaty of Paris in 1763, England was awarded all French territory east of the Mississippi River. The treaty, however, did not address Indian ownership. This was left to Sir William Johnson, King's agent for Indian affairs.

Johnson received mixed signals from London. Some members of the British government desired "Imperial regulation of Indian trade with restrictions of colonial expansion." Others "favored colonial expansion with colonial regulation of trade."

For five years, from the end of the French and Indian War to 1768, Johnson worked on establishing a new colonial frontier boundary line that would put "a limit to the westward movement of population...for preventing irregularities in the future and that the Indians may be convinced of our justice."

Johnson wanted to establish a western boundary line and "beyond this line no settlement was to be allowed until the whole Six Nations should think proper of selling part thereof."

More than 2,000 Indians and representatives of colonial America assembled at Fort Stanwix, near Rome, New York on November 5, 1768 to sign the Treaty of Ft. Stanwix. This treaty established a new American western border. Gifts and money totaling £10,460, 7 shillings, 3 pence were paid by the British government to seal the bargain.

John Penn, Governor of Pennsylvania and son of William Penn, attended the Ft. Stanwix Treaty sessions, and with the help of Johnson, was able to convince some of the Indians to resell their Wyoming lands to Pennsylvania. The Ft. Stanwix Treaty boundary line reflects this sale.

"The Proprietaries [of Pennsylvania] obtained a deed from the Six Nations, of the Susquehanna lands, being the same previously claimed to have been conveyed to the Susquehanna Company [of Connecticut]."

After this, there "commenced the strife, foot to foot, and hand to hand, of the conflicting parties, for the possession of this beautiful valley [that symbolized all of northeastern Pennsylvania]. The contest [is] still known as The First Pennymite <sic> War!"

Chapter 5

Great Secret of Northeastern Pennsylvania...

the black stones that burn - Anthracite Coal.

By meticulous attention to detail in 1768 at Ft. Stanwix (Rome, New York), the boundary line separating English/European settler from Indian gave all of the American Colonies' known mineral wealth to the white man.

Iron ore at Saugus, Massachusetts and western Connecticut, a copper processing mill in the Newark, New Jersey area, soft coal (bituminous) deposits in Virginia, a few gold nuggets from North Carolina, and the Old Mine Road from Kingston, New York to the copper mines at Pahaquarry (north of Delaware Water Gap on the New Jersey side of the Delaware River) were well known to the treaty makers in 1768.

But few knew about the black stones that burned.

The black stones that burn were unusual, but were they valuable?

Someone must have thought so, because when Ft. Stanwix Treaty was signed, the places where the stones were found, the great Wyoming Valley (Wilkes-Barre and Scranton) and the headwaters of the Lehigh and Schuylkill Rivers (Jim Thorpe to Pottsville), were in white man's territory.

The black stones lay hidden, north of the Blue Ridge mountain line, their secret known only to a few.

This was because of the geography of northeastern Pennsylvania: an area of "endless mountains" that could be penetrated only by three major river roadways.

Northeastern Pennsylvania was *Towamensing - the wild place* to the Indians.

It was *Saint Anthony's Wilderness* to the leader of the Moravian Christian missionaries, Count Nikolaus Ludwig von Zinzendorf.

It was thousands of acres of mountain land. From the Delaware River just one Indian trail cut westward through the "Shades of Death," where laurel and rhododendron flowers carpeted the forest floor like grass every June.

The *Minisink Path* connected the traveler with the Susquehanna River. It started at approximately today's Dingmans Ferry bridge and passed through Lords Valley, Blooming Grove, by Paupack Methodist Church, across the Wallenpaupack River to Hamlin, and from there entered the great Wyoming Valley through Cobb's Gap, picking up the Lackawanna River at Scranton. The Minisink Path then followed the Lackawanna to the Susquehanna River and down to Wyoming (Wilkes-Barre).

From Stroudsburg, another trail led directly to Wilkes-Barre, the *Pechoquealin Path.*

At Wilkes-Barre, the *Great Warriors Path*, coming from Shamokin (today's Sunbury) followed the Susquehanna River north into New York State where it connected with the *Forbidden Path.*

For travelers heading north from the Forks of the Delaware (Allentown, Bethlehem, Easton area) there were several routes to follow, depending on destination: the *Minsi/Delaware River*

Path followed the Delaware River to eventually connect with the Mohawk River near Albany, New York.

The *Lehigh Path* paralleled the Lehigh River northward and connected with the Susquehanna River south of Wilkes-Barre; the *Wyoming Path* was a direct overland route to Wilkes-Barre.

From Stroudsburg through Gnadenhutten (Jim Thorpe) to the Susquehanna River at Nescopeck (south of Wilkes-Barre), there was the *Nescopeck/Pohopoco Path*. At Hazleton, this path crossed the *Nanticoke Path* coming up from Reading.

For those who did not want to use the Great Warriors Path along the Susquehanna River, an overland route from Scranton, the *Lackawanna Path*, connected with the river at Great Bend on the Pennsylvania-New York border.

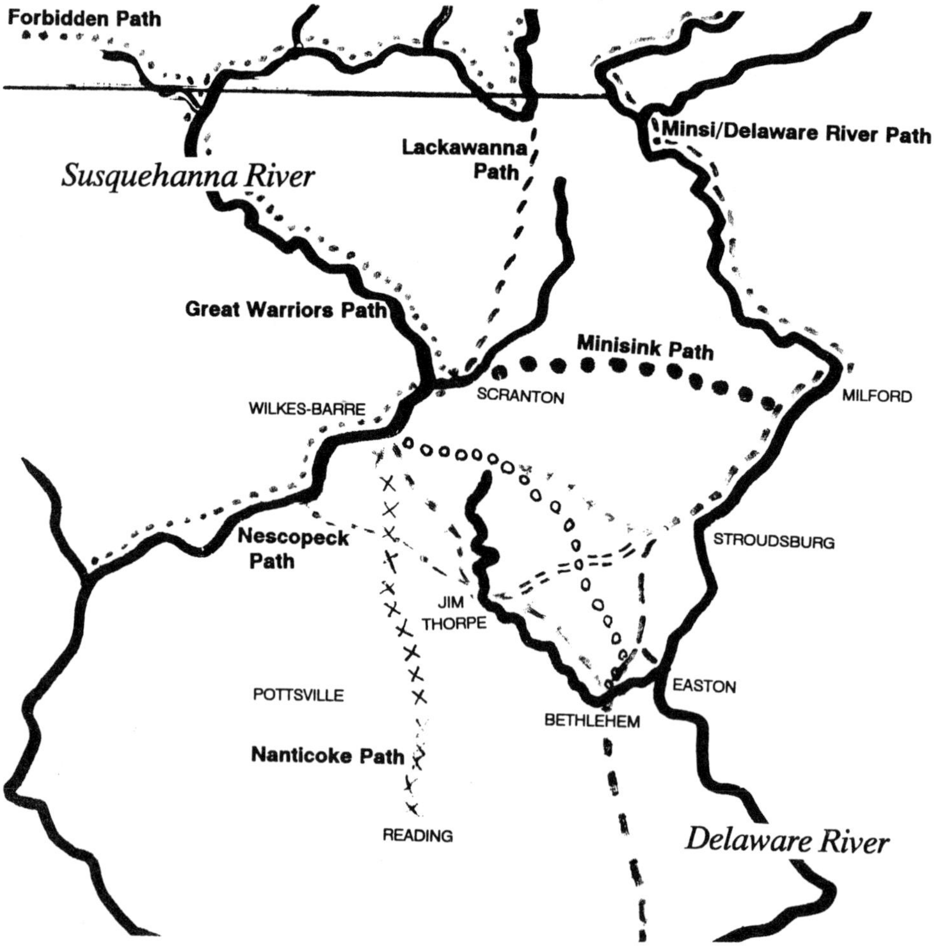

Lastly, a short path existed for the Indians to tend their apple orchard. From their encampment at the Lackawaxen and Delaware Rivers, the *Lackawaxen Path* trail took the Indians overland to present day Indian Orchard, south of Honesdale.

These were the paths through *Towamensing.*

No one lived in *Towamensing*.

Settlements were always along river banks, near the water necessary for life, and where animals came to drink.

The great secret of northeastern Pennsylvania would keep - to wait for a future day.

Chapter 6

Thus began the First Pennamite War.

"...and left the Susquehanna Company in undisturbed possession of the ground."

Immediately after the Ft. Stanwix Treaty in 1768, members of the Susquehanna Company met at Hartford, Connecticut and resolved "that five townships, five miles square, should be surveyed and granted, each to forty settlers...on condition that those settlers should remain upon the ground, 'man their rights' as was the phrase, and defend themselves and each other from the intrusion of all rival claimants."

The five townships in the heart of the great Wyoming valley, in the heart of northeastern Pennsylvania, were Wilkes-Barre, Hanover, Kingston, Plymouth, and Pittston.

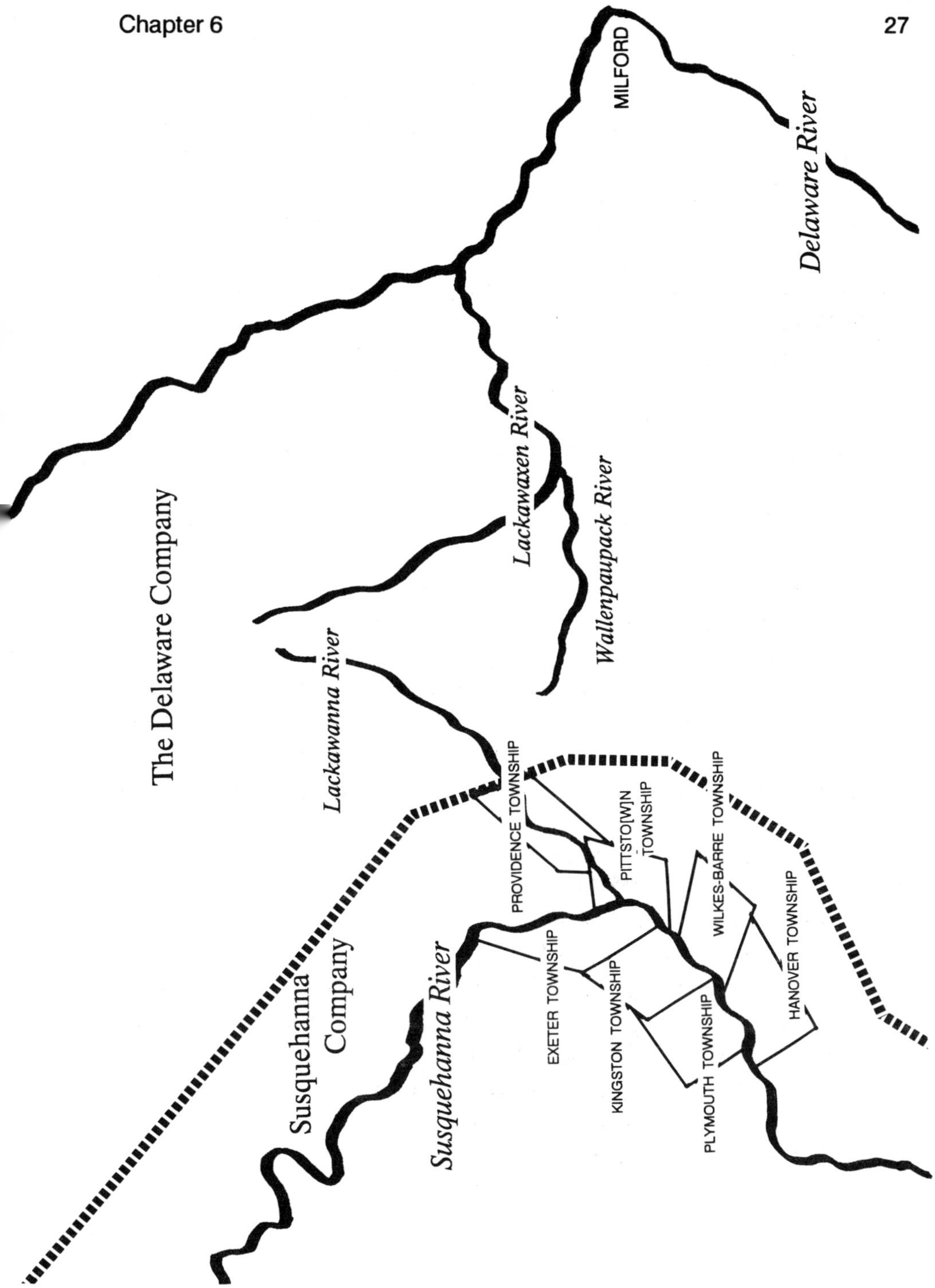
MILFORD
Delaware River
The Delaware Company
Lackawaxen River
Wallenpaupack River
Lackawanna River
PROVIDENCE TOWNSHIP
PITTSTO[W]N TOWNSHIP
WILKES-BARRE TOWNSHIP
HANOVER TOWNSHIP
EXETER TOWNSHIP
KINGSTON TOWNSHIP
PLYMOUTH TOWNSHIP
Susquehanna River
Susquehanna Company

(These townships later became districts with the same names before going on to become cities and towns, also of the same names.

Colonel John Durkee has always been acknowledged as the originator of "Wilkesbarre," as it was known at that time. The Yankee Durkee was an admirer, along with many other American colonists, of both John Wilkes and Isaac Barre, who were strong supporters of American rights in the British House of Commons.

Wyoming, an Indian word, sometimes interchanged with Wilkes-Barre by early settlers, can also refer to the Indian encampment on the Susquehanna River, or to a specific portion of the valley, to the entire valley, or to all of the Connecticut claim in north-eastern Pennsylvania.)

"On the other hand, the Proprietaries of Pennsylvania designated their leaders and marshalled their forces for the contest." Charles Stewart, Captain Amos Ogden and John Jennings, Esquire "constituted the Chief Executive Directory to conduct the Proprietaries' affairs at Wyoming [Wilkes-Barre]."

These three men were given a seven-year lease for 100 acres of land "upon condition that they should establish an Indian trading house thereon, and defend the valley from encroachment."

The express condition of defending was "the possession of these lands from the Connecticut claimants."

Stewart, Ogden and Jennings arrived at Wilkes-Barre in January, 1769. February 8 of the same year, 1769, forty Connecticut claimants arrived at Wilkes-Barre.

Thus began the First Pennamite War.

"In the bosom of the wilderness, far removed from any civilized settlement" the combatants were "fully aware that the eyes of anxious multitudes were upon them.

"All Connecticut was on tiptoe to watch, to cheer, and to sustain her adventurous colony at Wyoming."

Philadelphia, "gazing with anxious suspense, was equally eager, and not less zealously determined to stand by the party that had nobly volunteered to defend the honor and maintain the rights of Pennsylvania."

Both sides had their advantages. While Jennings, on the Pennsylvania side, was High Sheriff of Northampton County (county seat at Easton), "above the Blue Mountains, on the Delaware, a settlement existed nearby, or embracing Stroudsburg, the present seat of justice of Monroe County, whereof a number of the principal inhabitants united with the Connecticut people, and entered heart and hand into their cause."

Stroudsburg resident John McDowell, a Scotsman, became a true friend to the Connecticut Yankees; "he espoused their cause." (One of his daughters, Elizabeth, married Jacob Stroud.)

When the forty settlers from Connecticut arrived at Wilkes-Barre, they found the Pennsylvanians occupying the remains of their 1763 buildings.

Jennings, in his capacity as High Sheriff, soon seized three of the Connecticut settlers: Isaac Tripp, Benjamin Follett and Vine Elderkin.

"Gentlemen, in the name of the Commonwealth of Pennsylvania, you are my prisoners!" said Jennings; and the Pennsylvanians left with their prisoners for the county jail at Easton.

(At this time, before the Revolutionary War of 1776, from Pennsylvania's viewpoint all of Northeastern Pennsylvania was part of Northampton County with its county seat at Easton.)

Bail was soon raised among sympathizers at Stroudsburg and the three returned to Wilkes-Barre - followed by the Sheriff and a posse from Easton.

On the second time around, Jennings arrested almost all of the Connecticut settlers (about 30) at Wilkes-Barre. Once again bail was raised. On their release at Easton, the "prisoners" returned immediately to Wilkes-Barre, to be in quiet possession of the Wyoming Valley.

One month later, in April of 1769, the settlement at Wilkes-Barre numbered over 270 men!

"But no one supposed that peace and security were finally yielded them by their alert and powerful [Pennsylvania] opponents."

So while the Susquehanna Company shareholders went about their business of establishing homes, surveying boundary lines and planting crops, "every breeze from the southern mountain awakened fears of an approaching enemy."

In September 1769, Sheriff Jennings returned with 200 men, including an artillery company and a four pound cannon!

With "that terrible four pounder destroying every hope of victory," the Connecticut settlers vacated the valley and headed east.

But the Yankees came back to Wyoming in February of the next year, 1770. With the aid of Captain Lazarus Stewart, once leader of The Paxton Boys, Connecticut men not only regained the valley but also captured the dreaded cannon.

Pennsylvania Governor John Penn appealed to General Thomas Gage in mid-April of 1770 for aid. The General replied, "the affair in question seems to be a dispute concerning property, in which I cannot but think it would be highly improper for the King's troops to interfere."

Left on their own, Susquehanna Company shareholders triumphed once again over Captain Ogden and the Pennsylvanians. This time, they set fire to Ogden's headquarters, Ft. Ogden, taking the famous four pounder back to their own new fort, Fort Durkee.

Ogden returned September 21 to capture many Yankees, sending them to jail at Easton. Four Connecticut men attempted to reach those living at Coshutunk, on the Delaware River, for help, but Ogden caught them, too.

Defeated, the Connecticut settlers left Wyoming - again.

Then three months later on December 18, Captain Lazarus Stewart and 30 Yankees were back and in possession of Wyoming Valley!

But on January 20 of the next year, 1771, Ogden returned to Wilkes-Barre - and the Yankees found themselves marching once again to Easton, prisoners of the Pennsylvanians.

In control of the valley, Ogden built a new fort, Ft. Wyoming, just north of Connecticut's Durkee. Three months later, however, in April 1771, with Captain Stewart and Yankee Captain Zebulon Butler leading the way, Connecticut forces laid siege to Ft. Wyoming. To do so, they brought out of hiding the four pounder originally brought to the valley by Ogden.

Escaping the siege, Ogden reached Philadelphia at the end of July. There 300 pounds from the public treasury was used to raise a company headed by Captains Ogden and Dick.

Back in the valley, the Pennsylvania forces were ambushed. They rushed to Ft. Wyoming for safety but once inside, the gates were closed by the Connecticut men.

Two weeks later, on August 14, 1771, Ogden and Dick, along with Captain Morris, capitulated, and "all the Pennsylvania troops were forthwith to withdraw from Wyoming."

Captain Ledlie, rushing north to the aid of Ogden, Dick and Morris, received word in September to turn back.

"The Proprietary Government ordered the return of Captain Ledlie, and left the Susquehanna Company in undisturbed possession of the ground."

Chapter 7

"...to be called Westmoreland, attaching it to the county of Litchfield."

Sensing that they were truly in possession of the land, the Connecticut settlers made immediate plans to solidify their ownership.

They intensified settlement; they adjusted to handle, in an orderly fashion, dramatic population increases; and they formalized their activities.

One thing they never forgot - that they were pioneers living in a frontier settlement and had to make do as best as they could. Their lives depended on what they could do for themselves. There was a spirit of cooperation.

"New ground" had to be ready by July. Turnips were the first to be sown, followed by buckwheat. Artichokes and mustard were placed along the split rail fences at the corners. These were the crops that could be harvested immediately.

By every front door, a rosebush greeted the visitor. Near every house, an orchard supplied fruits. Flaxseed (for linen) and sheep (for wool) came with the settlers, too. The blacksmiths among them, using "stone coal," quickly forged spindles for spinning wheels and looms - otherwise no clothes!

For the first years, because the "road" from Delaware River westward was a "mere path through the woods," no furniture could be brought in by oxen or wagon. "Every man made his own or had none."

Mills were not built until later. So if grain or corn was to be ground, the people of the neighborhood would send a man on horseback, carrying as large a load of grain as possible, to the settlement at Coshutunk on the Delaware, where there was a mill. On the rider's return, the "fine flour" would be divided among the neighbors who had sent him.

The "fine flour" was saved for special occasions "to have something fine and rich for grand company and weddings."

These are the ingredients used for wedding cake by one of the early Connecticut settlers at Wyoming (Wilkes-Barre): 10 lb. flour; 11 lb. sugar; 12 lb. butter; 20 lb. currants; 4 lb. almonds; 4 lb. citron; 90 eggs; 1/2 lb. cinnamon; 3 oz. cloves; 2 oz. mace; 2 oz. nutmegs; 1 pt. brandy or coffee; and 1 pt. molasses.

Today's cook will note that the recipe did not include instructions for oven temperature or baking time.

The cake was put to bake in the fireplace oven until done. In small houses (they were always small at first), "an opening was cut through the logs in one end of the house large enough for the chimney and fireplace, and the chimney was then built on the outside tight up against the end of the house, closing the opening."

For visitors, one drink that was "always brought forth" was "metheglin," a fermented mixture of honey and cider or water. "From its mild and sweet taste it seemed that it must be innocent, but many have been deceived by it and knocked over."

For metheglin, "the crushed or squeezed honeycomb from all the [bee] hives taken up would be washed - sometimes with fresh new cider, but generally with water - and then melted and skimmed, and run into cakes for bee's-wax. The water or cider with which they washed the comb, was boiled and skimmed and put into a clean barrel and laid down in the [root] cellar to ferment. It was a long time fermenting, but after fermenting it was tightly bunged up and left to lie there undisturbed till the next summer at harvest time, when it was supposed to be fit for use."

Pork was a great staple, varied by fish, chicken, deer and game bird meats. Salt, for which there was no substitute, was very dear. It was used to preserve meats.

In the spring, everyone went down to the Susquehanna River for the Shad run.

In the summer, everyone worked.

With winter, more time was free for pleasures. "Everybody kept openhouse, as it were, even into the night, when sleigh-bells were heard in the crisp, cold winter air of the afternoon or evening." The sleighs would stop "where any young folk lived;" there "would be a rush of girls and boys with red noses and cheeks into the house;" the "metheglin, and mince pies, and donuts, and apples, and nuts were brought out and eaten and drank; and how they rushed off with shouts and laughter and tumbled into the sleigh or 'sled' among quilts and comforters and coverlets and straw and sped away laughing and singing and full of joyous noise to the next house."

This was living in the lands opened for development by the Susquehanna Company and The Delaware Company north of Latitude 41°.

But what was the life of the community north of Latitude 41°?

For one, there was no industry geared to export. There was nothing the Connecticut settlers were making that was wanted by the outside world. Whatever money did come into the area came primarily from peltry - fur pelts of animals.

The community then was entirely responsible for its own survival. First, Nathan Chapman, followed by Stephen Fuller, Obadiah Gore, Jr. and Seth Marvin, were given land and water rights to the stream where the Connecticut men had first settled, provided a grist mill and saw mills were built and ready for use no later than November 1, 1773. At the same time two ferries to service travelers crossing the Susquehanna were authorized.

The third action of the settlers was to send out a call for a clergyman to come and live among them. Fourth on the list was the decision "to raise three pence on the pound, on the district list, to keep a free school in the several school districts in the said Wilkesbarre."

Militia companies were established, and "any person selling liquor to an Indian was to forfeit his goods, and be expelled from the colony."

A land office was maintained where grants were surveyed and recorded.

Lastly, the Susquehanna Company, meeting at Hartford, Connecticut on June 2, 1773, adopted a government for the area. (The Delaware Company shareholders acted similarly.)

Three men in each community - Wilkes-Barre, Kingston, Plymouth, New Providence (Scranton), Pittston, and Hanover - were to act as Directors. From them, all local government operations would descend.

Dr. William Hooker Smith arrived in the valley in 1772. Later, several residents subscribed to bring in an additional medical man, Dr. John Caulkins, "in case he should settle among us in the quality of a physician."

Then, in January 1774, an act was passed by the General Assembly of Connecticut "erecting all the territory within her charter limits, from the river Delaware to a line fifteen miles west of the Susquehanna, into a town, with all the corporate powers of other towns in the colony, to be called Westmoreland, attaching it to the county of Litchfield."

No doubt about it, the land we know today as northeastern Pennsylvania was Connecticut land two hundred years ago!

Chapter 8

"A town meeting therefore, now, when 'legally warned,' called together all the Freemen, in all the townships or settlements, from the Delaware to fifteen miles beyond the Susquehanna, and from the Lehigh, north to Tioga Point."

When the Connecticut General Assembly approved Westmoreland Town (equivalent to a township) and attached it to Litchfield County, the northwesternmost county in Connecticut proper, two things were accomplished.

For the first time, shareholders of both the Susquehanna Company and The Delaware Company who were actually settling the "western lands" were officially recognized "as having the protection and being subject to the laws of the ancient and powerful government of Connecticut." Secondly, to obviate a possibly demeaning future, Connecticut was assuring its territorial lands originally given under royal charter.

The storm clouds of war were already on the horizon. Combatants and observers to the conflict in northeastern Pennsylvania had recognized for some time that Pennsylvania's seemingly "soft approach" to ousting the Connecticut claimants stemmed more from the Proprietors' preoccupation with the deteriorating relationship between England and her American Colonies than from military incompetency.

If war did break out, and if the Colonies were successful, Connecticut had assured her ownership of "western lands." If the American Colonies lost, Connecticut was still assured of ownership because of the Royal Charter of 1662.

So, "in accordance with the Act of Assembly, [Connecticut] Governor Turnbull issued his proclamation, forbidding any settlement within the limits of Westmoreland [today's northeastern Pennsylvania], except under the authority of Connecticut." To the Connecticut settlers at Wyoming and elsewhere between Latitude 41°N and 42°N west of Longititude 74°41' West, the Act of Assembly in January 1774 and Governor's Proclamation meant that they "could not be looked upon as a lawless band of freebooters."

When the Quaker, Richard Smith, made his famous trip in 1769, up the Hudson River and back down the Delaware River, from his home in Burlington, New Jersey, across from Philadelphia, he noted "16 or 17 farms" at Coshutunk (Damascus, Wayne County area). There, Smith hired the young Bazileel Tyler to guide his party down the Delaware.

"We saw but two small settlements between Coshutunk and the Minisinks" (Port Jervis, New York), Smith wrote in 1769. What houses were found were predominantly on the New York/New Jersey side of the Delaware River. (It was Peter Kikendahl who provided overnight lodging at Minisink/Port Jervis for the Smith party.) Leaving Kikendahl's for a 13 + hour downriver trip to near Easton, Smith noted, "Not many houses are to be seen..."

Eight years later, in 1774, it was another picture. Not only was the settlement at Coshutunk still thriving, but there was one on the Wallenpaupack River, too.

Several small settlements along the Delaware had also commenced. While the names of some of the settlers at Milford, such as Quick, are found first in records of New York State, and others, such as Wells, who operated a ferry, were known to be from Connecticut, the people experienced more Pennslvania domination than the settlers at Wyoming (Wilkes-Barre).Those living along the Delaware could easily be accessed by river travel, as they were directly north of Stroudsburg and Easton, where land titles were recorded under Pennsylvania authority. (Remember that Connecticut's claim extended south only to Latitude 41° North, thus excluding the Stroudsburg area at Lat. 40°59'N.)

As for the settlement on the Wallenpaupack, it had been renewed when a small group of Connecticut settlers, on their way to Wyoming (Wilkes-Barre), had come on the remains of the Carters at Paupack. Pausing, they decided not to continue but to erect their community at Paupack instead, in today's Pike County. Further west, along the Indian Minisink Path (which was "called the Old Wyoming Road by Connecticut travelers"), two Connecticut families, the Robert Strongs and the Jacob Stantons, settled at Little Meadows, south of today's Hamlin on Route 191 (in Wayne County).

No sooner had word of Connecticut Assembly's action erecting Westmoreland Town reached northeastern Pennsylvania than all of the inhabitants were "legally warned" (the equivalent of today's "legal notices" in newspapers' classified advertisement sections) that the first town meeting would be held March 1, 1774 at Wyoming (Wilkes-Barre).

"A town meeting therefore, now, when 'legally warned,' called together all the Freemen, in all the townships or settlements, from the Delaware to fifteen miles beyond the Susquehanna, and from the Lehigh, north to Tioga Point."

At the meeting, continued the next day (March 2, 1774), the old township designations from the first days of the Susquehanna and Delaware Companies were abolished. In their place geographical/political subdivisions called "districts" were established.

Some of the districts were essentially the same in size and name as under their old township name, i.e. - Wilkes-Barre, Hanover, Plymouth, Kingston, and Pittston districts.

However, the former Providence (Scranton) and Exeter communities, along with all the lands north and west to the township line (concurrent with New York boundary line), were now one district "by ye name of ye North District." For the settlers living in the area of the Wallenpaupack, it was voted, "And that Lackaway settlement and Blooming Grove, and Sheolah [Shohola], to be one district, and to be called by ye name of ye Lackaway District." Finally, "And that Coshutunk, and all ye settlements on Delaware, be one district...and known by ye name of ye East District."

In all, eight town meetings were held during 1774. At the legally warned meeting of November 22, 1774, it was voted that a committee "make inquiry into, and search after all persons that are suspected to have been taking land under the title of Pennsylvania, etc., and that they [the committee] have full power to expel any person or persons" who are judged unwholesome inhabitants because they are acting contrary to the proclamations of the Governor of the Colony of Connecticut. One man from each of the newly created nine districts served on the committee.

At earlier meetings during 1774, of the almost 100 men appointed to various public offices, eight were from the Lackaway settlement on the Wallenpaupack, and two from Coshutunk.

Captain Silas Parks, from Lackaway, wore several hats including justice of the peace until "It was discovered that he played cards. The intelligence was immediately forwarded to Connecticut, and he was superseded."

And at the October 17 meeting, it was voted that five men from the Wyoming Valley "be a committee to meet such gentlemen, as shall be appointed at or near Delaware, 'to mark out a road from that river to the Susquehanna'."

Lastly, a census taken in 1774 showed that there were 1,922 inhabitants in the new Town of Westmoreland.

Chapter 9

Pennsylvania turned its eyes eastward. "The dispute between Great Britain and the colonies, now approaching to an open rupture..."

And where was Pennsylvania while all this was going on in Westmoreland Town?

Other than to issue a corresponding proclamation from its governor, following that of Connecticut's Gov. Turnbull's, which "prohibited any settlement on the contested claims, under pretended grants from Connecticut, or any other than the authority of the Proprietaries,..."

Pennsylvania turned its eyes eastward. "The dispute between Great Britain and the colonies, now approaching to an open rupture, had already effected a sensible, and so far, not unfavorable influence on the settlement, as, by occupying the attention of the [Pennsylvania] Proprietary Government with more immediate affairs, the Connecticut people had been left undisturbed to extend and establish their possessions."

There was, however, something Pennsylvania did do. Very quietly, after a hiatus of 20 years, Pennsylvania erected three new counties between 1771 and 1773. One of the new counties was called "Westmoreland" (February 26, 1773), located in southwestern Pennsylvania. To all intents and purposes, it had Pittsburgh as its principal community.

Another, called Northumberland (March 12, 1772), divided Northeastern Pennsylvania into two counties - Northumberland and Northampton - the division line following today's Wayne and Monroe Counties' western border.

(The third new county was Bedford, March 9, 1771, in south central Pennsylvania.)

Eight more years would go by before Pennsylvania would approve another new county.

After Northumberland was erected, the Christian Indians far up the Susquehanna River at Wyalusing, near the New York state border (now Bradford County), hastened their plans to move to Ohio River country. They had been invited to settle there by some of the Delaware tribes.

The Christian Indian town, called Friedenshütten, was established by Moravian missionaries with permission of Togahaju, the Cayuga Sachem, ten years after the destruction of Gnadenhuetten in 1755 (near today's Jim Thorpe in Carbon County).

But after the Treaty of Ft. Stanwix in 1768, "even in Pennsylvania it had begun to be fashionable to break faith with the Indians...In addition to this, the controversy between Pennsylvania and the Connecticut people was beginning to assume a serious aspect..."

So a decision to move was made at the end of 1771. Three months after Northumberland County was established, on June 11, 1772, the Friedenshütten congregation "took up their march toward the setting sun."

One group "went by canoe down the Susquehanna and up the West Branch [of the Susquehanna River]. The [church] bell was taken down from its turret and carried by Anthony in his canoe in the van of the fleet, and was tolled until the squadron rounded the mountain a mile and a half below the church."

In the years following the closing of Friedenshütten, both Connecticut and Pennsylvania settlers moved into the area at a rapid rate. Many of the people were not in favor of independence; and still others were "disaffected people from the southeastern part of the province of New York." Together, they "became a dangerous element, ready to foment any disturbance which might distress the Whigs [political party in favor of American independence from England] and aid the cause of the British crown."

In the area immediately to the southeast, in what is now Wyoming County, Connecticut settlers were among the obvious majority. (Pennsylvania did not begin to send in settlers until 1773.)

The other northern lands (today's Susquehanna County) remained unspoiled, with no "white man" settlements before 1776.

And on the southern side of the new Northumberland County, the gradual expansion outward from Philadelphia had assured Pennsylvania title to properties.

The reasoning behind Pennsylvania establishing Northumberland County when it did is unclear, because the shift in Pennsylvania jurisdiction for the Connecticut-claimed area of Northeastern Pennsylvania was not actively noticed to the Yankee inhabitants. Supposedly, Pennsylvania was accommodating people who were visiting the area in search of land.

Northumberland was erected in 1772, but it was not until 1774 that Pennsylvania Assembly advanced a loan of £800 to Northumberland for building a courthouse and prison (the two main functions of county government) at the county seat, Sunbury.

And not until November 25, 1775, three years after establishing the new county, but within three months after defeating and destroying a Connecticut settlement near Sunbury, that was outside the boundaries of Connecticut's Westmoreland, would Pennsylvania Assembly request the Proprietory Governor "to give orders for a due execution of the laws of this Province in the counties of Northampton and Northumberland."

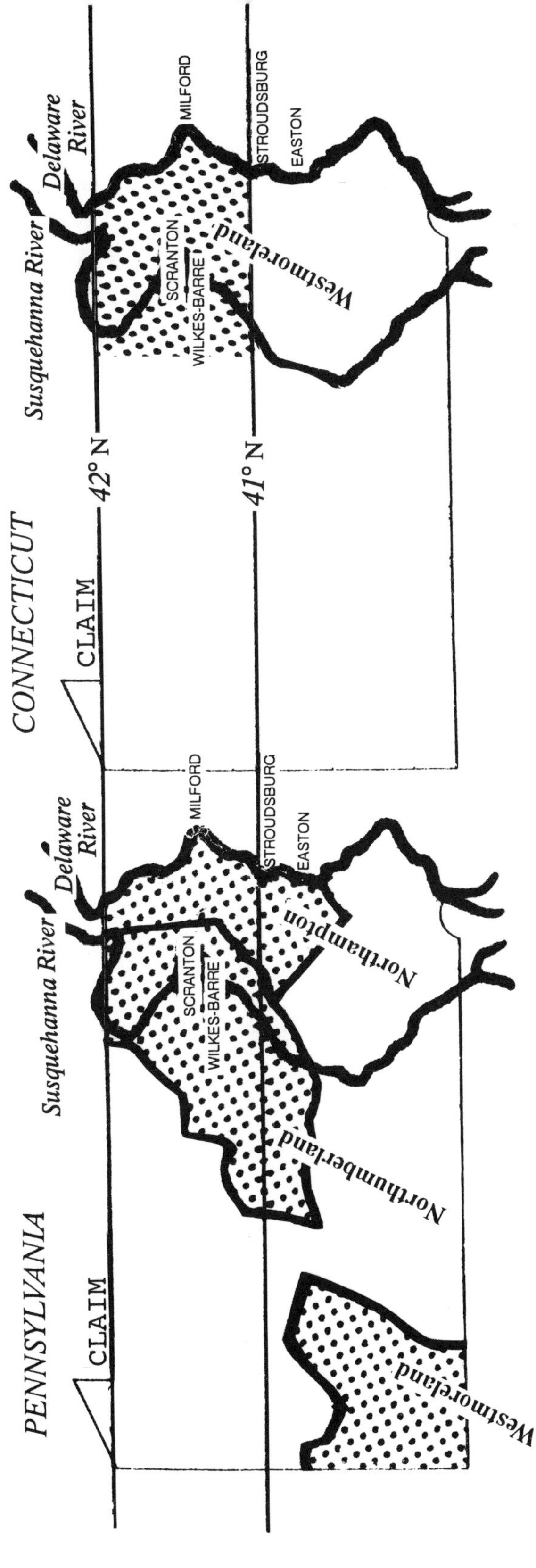
PENNSYLVANIA
CLAIM
Susquehanna River
Delaware River
MILFORD
STROUDSBURG
EASTON
SCRANTON
WILKES-BARRE
Northampton
Northumberland
Westmoreland
CONNECTICUT
CLAIM
42° N
41° N
Susquehanna River
Delaware River
MILFORD
STROUDSBURG
EASTON
SCRANTON
WILKES-BARRE
Westmoreland

The request and orders aimed directly at combating Connecticut's intrusion into Pennsylvania. *This is what Pennsylvania did, along with encouraging settlers to move into the northeastern area. Would it be sufficient? Or would northeastern Pennsylvania always be Connecticut-dominated?*

As for the reasoning behind Connecticut Assembly's designation of "Westmoreland" for the Susquehanna and Delaware Companies' lands, it also is 'foggy.' Pennsylvania's Westmoreland preceded Connecticut's Westmoreland by just eleven months.

But whatever Pennsylvania was doing didn't really matter to the Connecticut settlers between Latitudes 41° and 42° North and west of the Delaware River. They were too busy formulating their own Connecticut-style government.

Chapter 10

Four months later, on August 1, 1775, in the Town of Westmoreland, in the County of Litchfield, in the Colony of Connecticut, at a town meeting, legally warned, on the banks of the Susquehanna at a place called Wyoming (Wilkes-Barre), the residents voted, "that they will strictly observe and follow ye rules and regulations of ye Honorable Continental Congress, now sitting in Philadelphia."

Then, in January of 1776, Connecticut Assembly declared Westmoreland a separate county within Connecticut "with all the rights and privileges accorded every other county."

When Paul Revere made his midnight dash out of Boston to warn the residents of Concord, Massachusetts that the British were coming, the war between England and her American colonies began in earnest.

The Battle of Concord, Massachusetts, fought April 19, 1775 (giving victory to the British), marked the beginning of the military phase of the revolutionary era.

Four months later, on August 1, 1775, in the Town of Westmoreland, in the County of Litchfield, in the Colony of Connecticut, at a town meeting, legally warned, on the banks of the Susquehanna at a place called Wyoming (Wilkes-Barre), the residents voted, "that they will strictly observe and follow ye rules and regulations of ye Honorable Continental Congress, now sitting in Philadelphia."

In September 1775, the Connecticut settlers at Charleston and Judea, near today's Muncy, Pennsylvania, and east of the confluence of the East and West Branches of the Susquehanna River, were attacked. The settlements were destroyed by "Colonel Plunket, under orders from the [Pennsylvania] Government."

Charleston and Judea, part of Susquehanna Company lands, were west of the Westmoreland's boundary line set by Connecticut Assembly. With their destruction, Connecticut stopped legislation at the Westmoreland boundary.

However, Colonel William Plunket was not to be stopped. He and his troops pushed east, entering the Wyoming Valley at Nanticoke.

Scouts from Wilkes-Barre reported Plunket was coming up seven hundred strong.

"Seven hundred men! Nearly double the force Westmoreland could bring into the field.

"Of those who had taken the Freeman's oath, the whole number amounted to two hundred and eighty-five, and of these several came from the Lackawaxen settlement, forty miles east of Wyoming, a few from Coshutunk, on the Delaware, and many aged men were on the list."

Continental Congress, meeting in Philadelphia, heard of the impending battle and "in the better interest of the country as a whole" ordered on December 20, 1775:

"The Congress taking into consideration, the dispute between the people of Pennsylvania and Connecticut, on the waters of the Susquehanna, came to the following resolution:

"Resolved, That it is the opinion of this Congress, and it is accordingly recommended, that the contending parties immediately cease all hostilities, and avoid every appearance of force until the dispute can be legally decided."

Congress' ruling did not reach Plunket in time and he attacked.

"Colonel Zebulon Butler, who commanded the Yankees, by the most strenuous exertions had mustered about three hundred men and boys, but there were not enough guns to arm the whole." Many fashioned a weapon they called "the end of time" - a scythe fastened to a long handle, projecting straight out.

By Christmas Day, Plunket "concluded his rash enterprise by a retreat." He had found "Colonel Butler's position too strong to be carried by storm."

(The Plunket Invasion was never described as a Pennamite War event probably because Plunket was a known royalist. The Connecticut settlers designated the battle as an "invasion made by the Tory party of the Pennsylvania people.")

And because the conflict in Northeastern Pennsylvania "may be very prejudicial to the common interest of the United Colonies," Congress, on December 23, 1775, passed the following: "Whereas, the Colony of Connecticut has, by a certain Act of their Assembly, 'Resolved, that no further settlements be made on the lands disputed between them and Pennsylvania, without license from the said Assembly,'

"'Resolved, That it be recommended to the Colony of Connecticut not to introduce any settlers on the said lands, till the further order of this Congress, until the said dispute shall be settled.'"

With both Connecticut Assembly and the newly created Continental Congress freezing any further emigration for the remainder of the war years 1776-1782, population remained at approximately 2,500 in northeastern Pennsylvania.

And with the Connecticut people still "in possession of the ground."

Then, in January of 1776, Connecticut Assembly declared Westmoreland a separate county within Connecticut "with all the rights and privileges accorded every other county."

At the first town meeting legally warned following Connecticut Assembly's action, Westmoreland elected John Jenkins, Esquire, to represent it at the next session of Connecticut's General Assembly.

With an ever faster beat, political events began to shape Connecticut's Westmoreland County.

On July 4, 1776, the new United States of America declared its independence from England.

Over the summer, Westmoreland County established the 24th Regiment of Connecticut Militia. Additionally, about 30 men enlisted in the Continental Army, credited to Connecticut.

Then, on August 23, 1776, the new United States Congress sent word, "Resolved, That two companies on the Continental establishment, be raised in the town of Westmoreland, and stationed in proper places for the defense of the inhabitants of said town, and parts adjacent, till further order of Congress; the commissioned officers of the said two companies, to be immediately appointed by Congress."

The Resolution went on to state that the men would "be liable to serve in any part of the United States, when ordered by Congress."

In a little over a month, the able-bodied men of Westmoreland, Connecticut, had "flocked to the standard raised," and both companies were full, a total of about 165 men.

At the same time, it was formally agreed by all parties concerned (Connecticut, Pennsylvania, and U.S. Congress) that the question of land claims would be put aside until after the war, and that Connecticut settlers would remain in possession at least until then.

Far to the north, at Johnson Hall, north of the Mohawk River in New York State, Sir Guy Johnson and Sir John Johnson, successors to Sir William Johnson's position of King's agent, and his personal wealth and title, moved to consolidate all Indians on the side of the English.

Not to be outdone, Molly Brant, Sir William Johnson's Indian housekeeper, and mother of eight of his children, wept tears at "flaring council fires" and "was among [King] George III's best recruiting agents in America."

If the Indians fought, it would be for England!

Chapter 11

Meanwhile, to the west, at Wyoming (Wilkes-Barre), an important battle of the Revolution was shaping up. Afterwards, in history, it would be known as the Wyoming Massacre.

The Legacy left by Sir William Johnson, that of King's Agent, was not to be ignored. Those who had known him and worked with him during his lifetime strove to continue his world.

Johnson's nephew and later son-in-law, Sir Guy Johnson, succeeded Sir William as Indian Superintendent. A son, John, by Sir William's first German-born wife Catherine, inherited the baronet title and lands that went with the title. These two men, with another Johnson son-in-law, Daniel Claus, and the famed Indian who had been educated at The Reverend Eleazer Wheelock's Indian Charity School in Lebanon, Connecticut, Joseph Brant, directed Indian participation in the American Revolution War of 1776.

(Joseph Brant's parentage is obscure. It has been generally recognized that Molly Brant was his sister/step-sister, but whether or not Sir William was his father remains in doubt. Joseph had assumed the Brant surname.)

Other Johnson children, by his Indian wife, Caroline, and by his Indian housekeeper, Molly Brant, remained loyal to their father's support for King George III, too. It had been Molly's oldest son, Peter, who had captured Ethan Allen "with his own hand" at Montreal. And as the British attempted a two-prong attack to separate the New England colonies from New York, Pennsylvania and the southern colonies, it was Molly, herself, who had sent word to the British as they beseiged Ft. Stanwix at Rome, New York, of the advancing American relief forces headed by General Nicholas Herkimer. And to Joseph Brant had gone the title of "personal hero of the ambush at Oriskany," given when Herkimer died on the battlefield and it appeared the American-held Fort Stanwix might fall to the British.

Throughout these actions and all of the Revolution, the Iroquois Confederation walked a delicate tightrope of neutrality. But for the Indians who did fight, it was under the banner of English king and the successors to Sir William Johnson. From their vantage point on the north side of the Mohawk River about 40 miles west of Albany, the Johnson leaders worked tirelessly to keep the Indians on the British side.Molly, who had fled to Canada after the battle at Oriskany, "stoked Iroquois resentment against the patriots [Americans] who she felt had betrayed the policies of her man [Sir William]," by wanting to separate from England. At the same time, Sir John Johnson, "with Joseph Brant often serving under him as war chief, led pro-British Indians and the Scotch Highlanders who were his most faithful and martial tenants in raiding parties...The Mohawk Valley [of New York] became the Revolution's bloodiest cockpit."

Then in February of 1778, France signed a treaty of commerce and alliance with the new United States of America. This open support from a European country, traditionally an enemy of England, dramatically affected the war.

For one, British forces and Loyalists, learning that a French fleet was sailing to America, hastily evacuated Philadelphia to return to New York City. They feared the French would cut off their coastal supply line. Part of the transport movement back to New York was by boat. But for those British traveling overland across New Jersey, General George Washington and his troops overtook and attacked them on June 28, in what we know today as the Battle of Monmouth. Unfortunately, Washington was not able to prevail and the British successfully returned to New York.

Meanwhile, to the west, at Wyoming (Wilkes-Barre), an important battle of the Revolution was shaping up. Afterwards, in history, it would be known as the Wyoming Massacre.

A Connecticut man, John Butler, born at New London, Connecticut, had joined other ardent loyalists at the outbreak of the American Revolution. By May of 1776, Butler was deputy Indian agent under Guy Johnson. Later, Butler was assigned to Fort Niagara (at today's Buffalo, New York.)

Brant, like Butler, spent the winter of 1778 at Niagara. With spring, the Indian leader returned to the area south of the Mohawk River. It was his "intention only to gather provisions, and to get people for Butler's army and not fight or make any alarm." By April, Brant had gathered 1,500 fighters at an old Indian town on the upper Susquehanna River, at Unadilla, New York. Brant was "the most influential Mohawk of them all." South of Unadilla, another Indian town, Onaquaga, also on the Susquehanna, served as an additional gathering place. John Butler and his Rangers (American Loyalists or Tories) joined Brant the first week of June, 1778.

The combined force traveled down the Susquehanna. At Tioga Point (today's Athens, in Bradford County, Pennsylvania) Butler and Brant agreed that "Brant would continue collecting Tories and provisions." For his part, Butler would continue down the Susquehanna to his first major battle - at Wyoming (Wilkes-

Barre.) Later, after the Wyoming Massacre, Butler avowed "that it was his intention to harass the country to prevent them [Americans] from getting in their harvest. Schoharie [New York] and the Minisink [Delaware River area between today's Stroudsburg and Port Jervis], both of which had much corn and cattle, were his next goals, he boasted." Butler's plan was a good one, and it was viable. By attacking frontier settlements, Butler could cut back on American territory with little loss to his own troops. He did not fear Washington because he knew Washington would be too hard pressed confronting British soldiers to dispatch the necessary troops to defend isolated settlements.

There was another purpose to Butler's almost maniacal descent on Wyoming - a purpose unknown to Brant.

Soon after the new United States of America was formed on July 4, 1776, Matthias Hollenback, of Wilkes-Barre, began shipping the "black stones that burn" down the Susquehanna River to near Carlisle, Pennsylvania. "This fuel was hauled with great difficulty on water and overland to its destination, evidence of its esteem in the eyes of these patriot artisans." There, at the Carlisle Armory, "muskets, locks, sabres, as well as cannon forged from iron gads and hoops soldered together, were made during the war."

Hollenback procured the black stones from land he owned at Wyoming (Wilkes-Barre). Blacksmiths in Connecticut's Westmoreland County knew to use the anthracite in their forges when they wanted a more intense, longer lasting heat. Wilkes-Barre neighbors, brothers Obediah and Daniel Gore, had successfully experimented with the black stones at their blacksmith shop in the early 1770s.

But John Butler saw the black stones as a "secret weapon" that could win the war for the Patriots. Using "stones that burn," ironmongers could increase production of iron and cannon balls.

News of Butler's Rangers, from the time they first met with Brant at Unadilla, was carried to the settlers at Wyoming. Couriers also brought word of the gathering storm to Washington as he followed the British across New Jersey to Monmouth.

Chapter 12

Congress also "Resolved, that the two companies raised in Westmoreland be ordered to join General Washington WITH ALL POSSIBLE EXPEDITION."

But at home, back in northeastern Pennsylvania, few remained to defend.

When General George Washington learned of the gathering enemy forces high up on the Susquehanna River, he dispatched the American patriot, Colonel Zebulon Butler, instructing him to return to his home at Wyoming (Wilkes-Barre) and to strengthen defenses.

Zebulon Butler had come to Wyoming in the first wave of Connecticut settlers in 1769. Before that, he had "been much with British officers in the old French [and Indian] War, and with the accomplished French officers in the war of the Revolution, had a good deal of dignity and gravity about him."

Zebulon Butler was once described as "the gentleman [who stepped] from a door into the street and [met] a friend. With a slight bow, the gentleman [Butler], stopping, placed his silver headed cane under his left arm, took out his snuff box, rapped the lid as he took it off, and offered it to his friend; took a pinch himself, and brushing his face slightly with his doubled bandanna, bowed and passed on..." An officer and a gentleman, Butler would devote his life's work to the Wyoming Valley and to the new United States.

Butler returned to Wilkes-Barre at the end of June 1778 and "by common consent assumed the command of the Connecticut people."

In the intervening years, since August 1, 1775 when Westmoreland town had voted to follow the rules of the Continental Congress sitting at Philadelphia, Congress had made two requests to those living in northeastern Pennsylvania.

First, on August 23, 1776, Congress "Resolved, that two companies on the Continental establishment, be raised in the town of Westmoreland, and be stationed in proper places for the defense of the inhabitants of said town, and parts adjacent, till further notice of Congress; the commissioned officers of the said two companies, to be immediately appointed by Congress."

Three and a half months later, at their December 12, 1776 session, when Congress passed a resolution to adjourn to Baltimore City because "the movements of the enemy [British] have rendered this city [Philadelphia] the seat of war,"

Congress also "Resolved, that the two companies raised in Westmoreland be ordered to join General Washington WITH ALL POSSIBLE EXPEDITION."

So by the time John Butler and Joseph Brant met at Fort Niagara in the winter of 1778, the urgency of Washington's retreat to Valley Forge had removed the most able-bodied fighting men from defense of their homeland in northeastern Pennsylvania.

But apparently the Continental Congress's network of espionage and counter-espionage was more sophisticated than anyone realized, knowing where John Butler and Brant would strike. On March 16, 1778, Congress again "Resolved, That one full company of foot, be raised in the town of Westmoreland, on the east bank of the Susquehanna, for the defence of the said town, and the settlement on the frontiers, and in the neighborhood thereof, against the Indians and the enemies of these States; the said company to be enlisted to serve one year from the time of their enlisting, unless sooner discharged by Congress...That the company find their own arms, accoutrements, and blankets."

This now brought to three the number of fighting companies that Congress was asking of the residents of Northeastern Pennsylvania. The first two, raised in the later part of 1776, while always fighting as part of the Continental Army, were nevertheless identified separately, as "the two Independent Companies of Westmoreland." One company was headed by Captain Robert Durkee and the other by Captain Samuel Ransom. Together, the Independents "from Wyoming mustered about one hundred and sixty, and three hundred militia." (Zebulon Butler held his commission as regimental officer in the Connecticut line of the army.)

As for their fighting expertise, "the [Westmoreland] Companies were at Bound Brook, at Brandywine, at Germantown, and at Mud Fort;" had wintered at Valley Forge; and following the battle of Mill Stone River, near Somerset Court House, New Jersey, Zebulon Butler had received the following:

"Bound Brook, May 27th, 1777.

"Sir,- It is His Excellency, General Washington's orders, that you march immediately with the three detachments from Connecticut regiments, and the two companies of Wyoming men, to Chatham [New Jersey], there to take Gen. Stephens' orders, if there - if not, you will send to Head Quarters for directions."

So the fighting men of Westmoreland actively pursued the defense of their new country.

But at home, back in northeastern Pennsylvania, few remained to defend.

They realized their predicament, that they were vulnerable to attack from anyone coming down the Susquehanna River; "By detachments, the people worked on the several forts; built upon a larger scale, and with greater strength, but in the same manner as those of Forts Ogden and Durkee." The young, active men served as scouts "to guard the inhabitants from being surprised." The "old men formed themselves into companies, and performed duties in the forts. These companies of ancient men were called Reformados."

There had been special concern when a newly arrived family, by the name of Wintermoot, built their own private fort on the Susquehanna just north of the existing forts. By town meeting vote, two long time residents, the Hardings and the Jenkins, were asked to build their own fort to the north of Wintermoot's. This gave a total of five forts, in close proximity of each other at Wilkes-Barre, to stand in the place of the absent "Independent Companies of Westmoreland."

As a final show of strength, that famous four-pounder, originally brought to the Valley by Sheriff Jenning's Pennsylvania forces in 1769, was installed in Wilkes-Barre Fort. "Having no ball, it was kept as an alarm gun" - on hearing the 'boom' of the cannon, all those out in the fields were to rush immediately into a fort.

Colonel John Butler and his Rangers, with a detachment of Sir John Johnson's Royal Greens, almost seven hundred Indians, and Tories from New York, Pennsylvania, and New Jersey, left Tioga Point, on the Pennsylvania/New York border, and descended to where Bowman's Creek empties into the Susquehanna River, arriving there by the morning of June 30, 1778.

Scouting parties continued to bring word of the British/Indian/Tory advance down the Susquehanna River during June, 1778. "The people in the outer settlements fled to the forts." When word of developments at home reached the Independent Companies of Westmoreland, "the companies became nearly disorganized. Every commissioned officer but two resigned, and more than twenty-five of the men, with or without leave, left the ranks, and hastened to the Valley."

On June 23, 1778, Congress ordered the remaining men united into one company, and, by the Board of War, were apparently dispatched to Lancaster, Pennsylvania, where they were to march north to Wyoming (Wilkes-Barre).

The experienced fighting men, those who were ordered, and those who dared, were all on their way to the defense of north-eastern Pennsylvania.

Chapter 13

Wyoming's defenses were fast crumbling! "The Valley would be destroyed, piece-meal."

"Men, yonder is the enemy."

Then, "the battle being ended, the massacre began."

By the evening of June 30, 1778, eight of the residents at Fort Jenkins, the first line of defense for Wyoming against attack from the north, failed to return after a day of laboring in nearby fields. Later, when just John Harding returned, he described how, lying under the willows with only his mouth above the water of the Susquehanna River, he could hear "the dying groans of his friends," and the rustle of the attacking Indians as they searched the tall grass for him.

The next day, Colonel Zebulon Butler and his force marched from Forty Fort to the scene of the murders. "All were scalped and otherwise mutilated."

Immediately after Zebulon Butler and his men left the area near Jenkins Fort, to return to Forty Fort, John Butler and his Rangers, Indians and Tories entered the area from the north.

Butler's force took possession of the Wintermoot Fort without opposition. A visitor, Daniel Ingersoll and his wife, reported later that as they prepared to defend themselves as the Rangers approached, the Wintermoots indicated, "Colonel [John] Butler would be at home, there." The Ingersolls were made prisoners.

Wyoming's defenses were fast crumbling! "The Valley would be destroyed, piece-meal."

From Wintermoot's, John Butler sent Ingersoll, under escort, to Zebulon Butler, "demanding the surrender of Forty Fort and the Valley."

Again, on Friday, July 3, 1778, Ingersoll was sent with the message to surrender.

Afterwards, "Zebulon Butler called a council of war, and opinions were freely expressed." The majority opinion was that "to attack and defeat the enemy was the only hope of salvation for the settlement."

In the afternoon of July 3, "the column, consisting of about three hundred men, old men and boys, marched from the fort...In addition to those in the trainbands, the Judges of the Court, and all the civil officers who were near, went out."

The company at Pittston was left to guard the fort there, the company from Huntington and Salem area had not yet arrived; and the companies at Capouse (Scranton) and Lackaway settlement (Paupack, in today's Pike County) were too far away to render assistance.

It was left to "two hundred and thirty enrolled men, and seventy old people, boys, civil magistrates, and other volunteers" to mount the defense of Connecticut's Westmoreland County, that piece of land we think of today as Northeastern Pennsylvania.

A half-hour after the column had marched out of Forty Fort, Captain Durkee and Lieutenant Phineas Pierce arrived after riding all night. They were the vanguard of the consolidated Independent Company of Westmoreland that had marched to Lancaster, then north to "Reading, Bethlehem, Nazareth, the Wind-Gap, to Shoup's, four miles from Merwine's, where they were, on the 3rd of July - the day of the battle," still fifty miles away from Wyoming/Wilkes-Barre. (Today's Merwinsburg is on Route 115 about five miles north of Route 209.)

When Zebulon Butler and his band marched out of Forty Fort, John Butler's Rangers watched and reported on the advancing troop. The British Butler immediately ordered the torching of Wintermoot's Fort.

Just before the battle began, at approximately four in the afternoon of July 3rd, 1778, Colonel Zebulon Butler said,

"Men, yonder is the enemy.

"*The fate of the Hardings tells us what we have to expect if defeated. We come out to fight, not only for liberty, but for life itself, and what is dearer, to preserve our homes from conflagration; our women and children from the tomahawk. Stand firm the first shock, and the Indians will give way. Every man to his duty.*"

From the flaming embers of Fort Wintermoot to the banks of the Susquehanna River, the battle line was drawn: "On the American right was Captain Bidlack's company. Next was Captain Hewitt's, Daniel Gore being one of his Lieutenants. On the extreme left was Captain Whittlesey's. Colonel Butler, supported by Major John Garrett, commanded the right wing. Colonel Denison, supported by Lieutenant Colonel George Dorrance, commanded the left...such was the order of battle."

Facing the Americans was the British line. The left of the enemy, beginning at the smoldering fort, was commanded by Colonel John Butler himself. "A flanking party of Indian marksmen were concealed among some logs and bushes under the bank. Johnson's Royal Greens were next right and the main body of Indians formed the right wing. Everywhere, at every opening in between, additional Indian marksmen were stationed.

The two Butlers faced each other for battle.

Zebulon Butler gave the order to fire. He "ordered his men to fire, and at each discharge, to advance a step...As our men advanced, pouring in their platoon fires with great vivacity, the British line gave way, in spite of all their efforts to prevent it...As the battle waxed warmer, that fearful yell [from the Indians] was renewed again and again, with more and more spirit.

"For half an hour a hot fire had been given and sustained," then...

"The vastly superior numbers of the enemy began to develop its power." (Butler's Rangers and Indians were estimated at near 900, whereas the Americans were only 300.)

Colonel Denison gave the order for the left wing to "wheel back, so as to form an angle with the main line, and thus present his front, instead of flank, to the enemy."

As the manoeuver began, Indians "rushed in with horrid yells.

"Some had mistaken the order to fall back, as one to retreat, and that word, that fatal word, ran along the line.

"Utter confusion now prevailed on the left.

"Seeing the disorder, and his own men beginning to give way, Colonel Zebulon Butler threw himself between the fires of the opposing ranks, and rode up and down the line in the most reckless exposure.

"*'Don't leave me, my children, and the victory is ours!'*

"*But it was too late.*"

Soon the Indians were sixty rods *behind* the American line!

On the right, Captain Hewitt was asked, "*Shall we retreat?*"

"*I'll be damned if I do!*" was Hewitt's answer. "*'Drummer, strike up,'* cried he, and strove to rally his men. Every effort was vain. Thus he fought, and there he fell!"

Then, "the battle being ended, the massacre began."

Chapter 14

"It was a dreadful hour; men seemed transformed into demons!"

After the Indians broke through the American/Yankee line and found themselves about 1,000 feet behind the Wyoming defenders, they rushed to seal off retreat. The Connecticut men could escape only by the Susquehanna River.

"Those who remained near the line of battle saw the main body of the Indians hastening after the fugitives."

It would be sometime before concealing darkness would afford the losers protection for their movements. Many of the Yankees hid in the tall grass of their fields of grain that they had sown in the spring. Others sought refuge along the river bank, or if possible, would swim to an island in the Susquehanna.

Back "at Forty Fort, the bank of the river was lined by anxious wives and mothers, awaiting the issue." As the sound of gun shots became more infrequent, and nearer, "hope sank in dismay."

The long daylight served only to give the British John Butler and his men more time to seek out their defeated foe and take them prisoner - or worse, inflict mortal wounds. "It was a dreadful hour; men seemed transformed into demons!" With night, the situation worsened. "Whatever may have been the motive, the vast depths of hell, boiling with demoniac passions, never could have devised or executed such horrid tortures, as many of the Connecticut prisoners were that night doomed to endure." The British "Captain John Butler, much agitated, as the peculiar affluvium of burning human flesh came to his nostrils, said, in the hearing of Mr. Ingersoll, 'It is not in my power to help it.'"

The massacre lasted all night.

For the Connecticut settlers, "The night throughout the Valley was one of inexpressible anguish and despair."

Of the band of 300 that had marched from Forty Fort at three in the afternoon of July 3rd, 160 fell in battle or were taken prisoner and subjected to death by torture. Colonel Zebulon Butler reached Wilkes-Barre Fort, "and cast himself exhausted on the ground." Ensign Matthias Hollenback swam to Monocacy Island and from there escaped to ride all night to Captain Spalding's position near Merwine's (in today's Monroe County) "to apprise him of the state of affairs at Wyoming." Colonel Nathan Denison gathered the few soldiers who had come in to Forty Fort and stationed them to protect the women and children there.

On the morning of July 4th, 1778, precisely two years after the United States Declaration of Independence, all seemed lost for the new country's western frontier. "*The only hope of safety seemed to be in flight.*" The way east, through the 'Shades of Death' to the Delaware River, to Stroudsburg, and to Wind Gap, was crowded with families of women and children accompanied by just one or two men. Anderson Dana's widow (he had died on the battlefield) filled a pillow case with her husband's papers before leaving for Connecticut. (These papers preserved land and other records.)

Under John Butler's proposed terms of surrender, Zebulon Butler and those remaining of "Hewett's company [14], being continental soldiers, should be surrendered prisoners of war." Colonel Denison, negotiating with the British Butler, left to consult with Zebulon. There, the decision was quickly reached that the intended prisoners should leave the Valley immediately, to be out of reach of Butler's Rangers.

Denison returned to the British Butler to say that the men were no longer in the area. "Terms of capitulation were agreed upon...At the appointed time, the victors approached [Forty Fort] with colors flying and music playing.

"On paper the terms are fair, and security promised by the victor to life and property -

"But the Indians began to rob and burn, plunder and destroy..."

Later, for a third time, Colonel Denison sent for John Butler to complain of the injustices. "Colonel Butler," Denison said, "we have surrendered our fort and arms to you on the pledge of your faith that life and property should be protected. Articles of capitulation are considered as sacred by all civilized people." The British Butler vehemently complained that he could "do nothing with them."

On Wednesday, July 8th, John Butler and his Rangers departed from Wyoming - *turning north, back to Indian land!*

Even so, Connecticut settlers, the residents of Westmoreland County, from the Delaware River to 15 miles west of the Susquehanna River, knew no peace. One by one, individual settlements were destroyed in the days that followed. The residents of the small settlement at Little Meadows (near today's Hamlin, Wayne County) were wiped out; and those at Wallenpaupack (in Pike County) learned of the tragic battle at Wyoming when a young man fleeing east brought word of the fight and its outcome.

Even though the Paupack settlers had been benefiting from Joseph Brant's protectionist orders because the Wallenpaupack people "had always been kind to the Indians," they decided to leave.

They had reached the Delaware River about three miles above Milford on the evening of July 5th, when they learned that their homes and fort had been plundered, but supposedly by a band of outcast Indians living on the eastern side of the Delaware River (in Sullivan County, New York) near Cochecton.

As for Joseph Brant, when the Indian leader, who had once been entertained by King George III, left John Butler and his band of Rangers/Indians/Tories at Tioga Point (Athens, Pennsylvania) just prior to the Wyoming battle July 3rd, 1778, he headed north and east, returning to his Mohawk homeland. By August 29, 1778, Brant had again assembled another 600 Indians and 500 Loyalists/Tories at Onaquaga and Unadilla.

Brant's command position at Onaquaga (approximately 15 miles east of Binghamton, New York), situated on the upper Susquehanna River and close to the Delaware River, worried New York's Governor George Clinton. "'..the principal Place for Rendevous for the Enemy' should be destroyed," he wrote to General George Washington in October of 1778. But even as Clinton was writing, another Butler, this one named William, one of four military brothers from Carlisle, Pennsylvania who were all Patriots, was destroying not only Onaquaga but also Unadilla during an absence of Brant.

And because John Butler and his Rangers had returned north, vacating Pennsylvania's frontier, Zebulon Butler and 60 men, including 14 from Hewitt's Company, were able to regain possession of Wyoming, establishing headquarters at Wilkes-Barre by August 1, 1778. Colonel Thomas Hartley and the 11th Pennsylvania Regiment next made headquarters at Fort Muncy (near today's Williamsport on the West Branch of the Susquehanna River), and the other American Butler, William, at the head of the 4th Pennsylvania Regiment and a detachment of Morgan's Riflemen, set up headquarters at Schoharie Fort (25 miles due west of Albany, New York). In the Continental Army's effort to move against the enemy on the western frontier, their mission was to secure the frontier and rendezvous at Tioga Point.

Chapter 15

Sullivan's March ended the threat of a widespread invasion by the British and Indians from the northwest.

The three American Patriot leaders were unable to keep their appointment with destiny in the fall of 1778.

Colonel Zebulon Butler and his men at Wyoming, numbering about 60, and Colonel Thomas Hartley and the 11th Pennsylvania Regiment, numbering about 200, of which 130 were Wyoming men, did sweep north as far as Tioga Point (Athens in Bradford County, Pennsylvania). Their joint campaign achieved little because the British and Indian forces melted away into the woods rather than engage in a battle. Hartley and Butler did not continue pushing northeast to meet up with Colonel William Butler. Instead, they destroyed four Indian towns along the Susquehanna River between Tioga and Wyalusing before returning to their headquarters.

The same was true for the third member of the expedition, Colonel William Butler. He and his men, the 4th Pennsylvania

Regiment and a detachment of Morgan's Riflemen, numbering about 300 in all, after occupying Schoharie Fort, moved down the upper Susquehanna River to destroy the British-allied enemies at Unadilla and Onaquaga, New York. But Joseph Brant and his Indian force always seemed to be somewhere else whenever William Butler attacked.

So, like Hartley and Zebulon Butler, William Butler returned to his original base to await reinforcements. There would be no invasion of Indian and British territory at this time.

For Brant, the destruction of his old home area, Unadilla/Onaquaga, marked a change in his participation in the Revolution War of 1776.

On November 12, 1778, Brant and his Indians, with some of the British John Butler's Rangers and Tories headed by John Butler's son, Walter Butler (How's that for confusion - we now have four Butlers, of whom only two are directly related, running around in the woods chasing each other?), descended on Cherry Valley and massacred the inhabitants! (Cherry Valley is about 15 miles northeast of Cooperstown/Lake Otsego, New York.)

A month later, Brant wrote to an American officer: "..your Rables came to Onghquago [Onaquaga] when we Indians were gone from our place, and you Burned our Houses, which makes us and our [Iroquois] Brothers the Seneca Indians angrey, so that we Destroyed men, women and Children at Chervalle [Cherry Valley]." (Brant was a Mohawk, another of the six Iroquois nations.)

For the remainder of 1778 and the winter months of 1779, the British Butler's Rangers and Tories and the Indians under Joseph Brant controlled the western frontier and the lands beyond.

The Commander-in-Chief of the American forces, George Washington, recognized the seriousness of the situation - and that something would have to be done. Hartley, Zebulon Butler and William Butler had had to pull back from the invasion of the British/Indian strongholds, because they had not had enough fighting men.

Washington prepared to assemble a fighting force second to none.

It would be a monumental task to transport and supply approximately 3,500 soldiers in a march from the Delaware River to Niagara Falls.

General Washington selected General John Sullivan to lead the campaign, and General James Clinton, brother of New York Governor George Clinton, to join forces with Sullivan.

On March 22, 1779, Washington also dispatched Colonel Daniel Brodhead (of Brodheadsville, in Monroe County) to Fort Pitt (now Pittsburgh). Brodhead was to coordinate with Sullivan and Clinton.

Washington directed Sullivan to assemble his forces at Easton, Pennsylvania. Hartley, Zebulon Butler and Wyoming men were to serve under Sullivan.

Clinton was directed to bring William Butler's 4th Pennsylvania and detachments from Morgan's Rifle Corps, as well as the 3rd, 4th, and 5th New York and 6th Massachussetts Regiments, detachments from Lamb's Artillery, Captain John Harper's Volunteers, and Lieutenant Colonel Albert Pawling's Levies down the upper Susquehanna River to join Sullivan at Tioga Point on the New York-Pennsylvania border.

Throughout the weeks of May, June, and July of 1779, men and supplies were assembling. Brant kept watch on the activity. Then, in late July, 1779, Brant descended the Delaware River and struck at the Minisink region. A battle was fought on July 22. Later, Brant wrote in a letter, "We have burnt all the settlement called Minisink except one fort..." The next day, after marching fifteen miles, Brant and his men fought the enemy on the east (New York) side of the Delaware River, in the town of Tusten (Sullivan County) across from the Lackawaxen River where it empties into the Delaware in Shohola Township, Pike County.

Brant concluded his letter, written July 29, with, "We have taken forty odd scalps, and one prisoner, a captain. I suppose the enemy have lost near half of their men and most of their officers. They all belonged to a militia and were about 150 in number." (The battle was so fierce and so devastated the countryside that it was not until 43 years later in 1822 that the bones of the slain were gathered from the battlefield and given burial.)

The Battle of Minisink, however, did not divert General Washington from his schedule for the planned invasion of Indian/British Territory.

Sullivan, who had the most difficult route because it was overland with no roads wide enough for wagons, was already on his way. Washington had ordered him to start out from Easton on June 18, 1779.

On August 5, 1779, Washington instructed Clinton, at the Schoharie Fort, to proceed down the Susquehanna, beginning August 9; Brodhead, in the far west, was to take his more than 600 men up the Allegheny River starting August 11.

Leaving Easton, Sullivan crossed the Blue Mountain ridge at Wind Gap (in Schuylkill County), and headed for Learn's Tavern (on Route 611 in Tannersville, Monroe County), the last house on Pennsylvania's frontier. Traveling further north, at the top of the Pocono Plateau (Rt. 611, Tobyhanna Township, Monroe County), Sullivan and his men camped. The General named the swamp Hell's Kitchen and the area Hungary Hill.

From there, Sullivan forces reached the Susquehanna River, marching to Wilkes-Barre and the site of the Wyoming Massacre. As Sullivan proceeded up the Susquehanna, Clinton proceeded down the river.

On Thursday, August 19, 1779, Sullivan, Clinton, and their troops met about five miles west of Binghamton, New York. The combined forces then followed the Susquehanna west to Athens, at which point the soldiers turned northwest to follow the Chemung River, going towards Elmira, New York. (In 1779, Elmira was known as Newtown, a principal city of the Seneca Indians.)

On August 29, only Joseph Brant and that portion of Butler's Rangers commanded by the son, Walter Butler, were present to stand against Sullivan's men at Newtown. (*Today*, New York Route

17 runs through the middle of the battlefield.) The British at Fort Niagara had been unable to believe Brant's report that a force of 3,000 Americans were on the march in Indian territory; no reinforcements were sent.

Brant had no help. After the battle had been lost to the Americans, he and his men again faded into the woods.

Sullivan's March ended the threat of a widespread invasion by the British and Indians from the northwest.

Chapter 16

"We are also unanimously of opinion, that the jurisdiction and pre-emption of all territory lying within the charter of Pennsylvania, and now claimed by the State of Connecticut, do of right belong to the State of Pennsylvania."

When General John Sullivan and his men returned to Easton on October 15, 1779, a day of thanksgiving was proclaimed for October 17. Afterwards, the "victorious little army separated, taking up new positions assigned by his Excellency the Commander in Chief," General George Washington.

Colonel Daniel Brodhead, who had headed a simultanous expedition up the Allegheny River (above Pittsburgh) also received recognition for his efforts. Brodhead and his men had ascended more than two hundred miles up the Allegheny. Like Sullivan, Brodhead had fought a victorious battle, afterwhich the Indians disappeared into the woods.

Amidst all the hoopla of toasting the victorious Sullivan and Brodhead, the Pennsylvania Legislature quietly approved a bill on November 27, 1779 "for vesting the Estates of the late Proprietaries of Pennsylvania."

By this Act, the heirs of William Penn lost all ownership over Pennsylvania soil except for those manors surveyed and recorded prior to July 4, 1776. A sum of £130,000 Sterling was set aside by the Legislature as compensation, payment to Penn heirs to begin at the close of the war. People who had purchased land from the Penns were automatically recognized as having valid title to their properties. The remainder of Pennsylvania became "vacant land" the title resting with the Commonwealth.

As for Sullivan's campaign, it effectively stopped any large scale invasion by British and Indian forces into the heavily populated areas of New York and Pennsylvania. It did not, however, bring to an end all activities by the enemy.

The Connecticut settlers of Northeastern Pennsylvania continued to suffer depredations and killings throughout the rest of the war years. Those who had fled eastward in the summer of 1778 after the Wyoming Massacre attempted to return - to bury their dead, to harvest their crops, to protect their wordly goods, their houses.

But they found that they were always at the mercy of small, marauding bands of Tories and Indians. The killings continued.

The situation was so serious that on June 23, 1780, the Connecticut General Assembly passed a resolution abating all Westmoreland County taxes not yet paid, "to be considered as in part compensation for their losses," in conjunction with compensation coming from the new United States.

Only those along the Delaware River, the inhabitants of the first Connecticut settlement at Coshutunk, seemed to be free from danger.

On that fateful day, July 22, 1779, when Joseph Brant and his men had rampaged throughout the Minisink, "every man capable of bearing arms" had rushed to the Delaware River to take part in the Battle of Minisink.

Afterwards, the few remaining settlers expected to be destroyed. But the Indians who were seen did little damage. Gradually Coshutunk began to grow and flourish.

So, by 1780, in all of Connecticut's Westmoreland County life was beginning to assume some resemblance to normality. Even Matthias Hollenback had a chance to ship two more loads of anthracite, the black stones that burn, from his mine at Wilkes-Barre down the Susquehanna River to the Armory at Carlisle, Pennsylvania. This time the anthracite coal was used to fire forges for making muskets.

But there were still many hardships. In addition to enemy raiding parties, procuring basic necessities was a problem, as none of the mills for grinding flour remained. In August of 1780, several men "set out to Col. Stroud's to mill." (This was a fifty mile trip, one way, to Stroudsburg.)

Only fish as a food supply seemed to be plentiful. During the months of September, October, and November 1780, over 14,000 eels were taken as they descended the Susquehanna River.

However by the summer of 1781, the war for American independence seemed to have taken an alarming turn. England's Lord Charles Cornwallis, who had won the Battle of Brandywine and occupied Philadelphia, was directed to blockade the southern states. Starting in May of 1780, Cornwallis moved through first South Carolina then North Carolina and on into Virginia, conquering everywhere he went.

At Yorktown, Virginia, fifteen miles due east of Williamsburg, General George Washington engaged Cornwallis in a battle, aided by the French fleet.

On October 19, 1781, Cornwallis surrended to Washington!

Victory for the new nation, the United States of America, was near.

"Almost immediately after Lord Cornwallis' surrender, the contest [between England and America] being regarded as virtually closed, that is, on the 3rd of November..a petition was presented to Congress, 'from the Supreme Executive Council of Pennsylvania, stating a matter in dispute between the said state [Pennsylvania] and the State of Connecticut, respecting sundry lands lying on the east branch of the river Susquehanna, and praying a hearing in the premises, agreeable to the ninth article of the Confederation.'"

For the remainder of the year 1781 and until November 12, 1782, both Connecticut and Pennsylvania prepared for the hearing. Five commissioners, from New Hampshire, Rhode Island, Virginia, and two from New Jersey, opened their court at Trenton, New Jersey on November 12. The hearing would last forty-one judicial days.

On Monday, December 30, 1782, the Decree of Trenton was issued:

"We are unanimously of opinion that Connecticut has no right to the lands in controversy.

"We are also unanimously of opinion, that the jurisdiction and pre-emption of all territory lying within the charter of Pennsylvania, and now claimed by the State of Connecticut, do of right belong to the State of Pennsylvania."

"With the close of the year 1782, and the Trenton decree, the jurisdiction of Connecticut, which had continued nine years [over northeastern Pennsylvania], ceased;" there were no more town meetings. The old parchment bound book which had recorded the transactions of the Connecticut settlers, from the Delaware River to 15 miles west of the Susquehanna River between Latitudes 41° and 42° North, was closed. *...But was it the end?*

Chapter 17

"We do not think the lawful defense of what we esteem to be our own, can with any justice be termed a disaffection to Government..."

Once the Decree of Trenton was issued and Connecticut withdrew from the scene, the settlers at Wyoming yielded obedience to the laws of Pennsylvania. But the question of "Right of Soil" remained to be answered.

The commission, sitting at Trenton, New Jersey in December 1782, understood that they were empowered only to answer the *question of jurisdiction* over a specific land area. The commission knew it did not have the power to determine individual private land ownership, known as "Right of Soil."

And the Connecticut settlers, in becoming Pennsylvanians, still wanted to keep the parcels of land each had bought with his own money, his Right of Soil.

It was at this point in time that the territory thought of as northeastern Pennsylvania took on a new shape, becoming a circle with its center at Wilkes-Barre. ...WHY?

The Great Secret of Northeastern Pennsylvania was no longer a secret! And the land grab was on...for wherever the black stones might be found. The shipments of "the black stones that burn," by Matthias Hollenback from his farm at Wilkes-Barre, to the Armory forge at Carlisle in 1776 and again in 1780, had shown their potential value.

When Americans realized, after Cornwallis' defeat at Yorktown, Virginia October 19, 1781, that victory would be theirs sooner or later, all of the states and their inhabitants began positioning themselves for their future role in the new country. Pennsylvania put itself among the first to move in that direction when it petitioned the Continental Congress on November 3, 1781 for Connecticut's Westmoreland County.

Of all the new states in the new union - the United States of America - Pennsylvania wanted to possess the greatest natural treasures if it could. The state knew it already had a limitless supply of trees for timbering, soft coal (bituminous) deposits near Pittsburgh, and all four types of iron ore in abundant quantities. If it could also own the unusual "black stones that burn" whose worth had already been shown, then Pennsylvania's premier position as the keystone state was assured.

No nation can hope to succeed as a respected member of the international community if it cannot produce that most valuable of all commodities: iron. The United States of America knew that it would need the capability to produce iron and iron products.

Iron and its alloys are so valuable because it is iron that becomes a tool, a knife, a cooking pot, a weapon, or a structural support for a building. Not only should a country be able to supply its own demand for iron, it should be able to produce enough for export, too. In the western world, iron technology dates from about 1200 B.C., although Egyptians were known to have had iron ornaments prior to 3000 B.C. In America, iron ore was first discovered in

North Carolina in 1585 by an expedition sponsored by Sir Walter Raleigh. The English planned to manufacture it on site, as there was an abundance of trees for firewood. Trees were not too plentiful in England.

Before the American Revolution in 1776, the colonies had, at one time, produced as much as one-seventh of the world's iron supply, shipping it all to England for fabrication.

An iron furnace melts raw iron ore, pouring the iron into a mold. The mold can be for a specific object, such as a pot; or it can be for a bar of "pig iron." At an iron forge, the "bar" is softened and reshaped by hammering, or remelted and poured into a specific-shape mold.

Iron furnaces and iron forges both require very high temperatures. Sometimes temperatures as high as 3000° F must be sustained for many hours. In colonial America, wood in the form of charcoal was used to start a blast in a furnace. Some furnaces, once started, would operate continously for a year before shutting down. A rough estimate has a furnace using an acre of wood (charcoal) for every day it is in operation.

With an iron furnace's almost insatiable need for wood, ironmasters located on "plantations," anywhere from 375 acres to 10,124 acres in size, in the vicinity of iron ore deposits. From 1716 until 1783, when the American Revolution was officially ended by the Treaty of Paris, 84 ironworks operated in Pennsylvania. All but two were south of the Blue Ridge Mountain. Of the two, just one was in Connecticut's Westmoreland County: M. F. Alden began operation of Nanticoke Bloomery in 1779 on Nanticoke Creek. (A bloomery melts only small quantities of iron ore.)

So after Hollenback shipped the black stones to Carlisle Armory, the great secret in iron making became known to many outside Westmoreland: throwing a few of the "black stones that burn" on the fire produced a better heat; and one ton of the black stones was equal to 200 bushels of charcoal, a savings of a quarter of an acre of wood and the man hours needed to cut and char the wood, a task requiring anywhere from three to ten days.

Then, in 1783, the year the American Revolution was officially over, "the invention of the puddling furnace made it possible to melt pig iron under an anthracite ['the black stones that burn'] fire and stir the molten mass until it became wrought iron."The puddling furnace eliminated "finery" and "chaffery" operations, the pounding by tilt-hammers to reshape "blooms" into one-inch thick iron bars for reworking into horseshoe and hinge material."The puddling furnace rendered obsolete the colonial forge and its many stages of pounding and refining that was in the 18th century the only known method of producing workable wrought iron."

At last, "the black stones that burn" had a value! Pennsylvania could give America the capability it needed - iron and hot heat.

Johann D. Schoepf, a German natural scientist traveling in the United States in 1783 and 1784, wrote that "people in the Lehigh Valley [in Pennsylvania] continuously dreamed of mines and sudden wealth."

But no one had to dream of finding mines and sudden wealth if they owned land in the Wyoming Valley. Outcroppings of the black stones were everywhere in the valley.

So, on April 22, 1783, the Committee of Pennsylvania Landholders directed a commission appointed by the Pennsylvania General Assembly to inform the Connecticut settlers that they had one year in which to give up their land. Widows of husbands killed by Indians were allowed an extra year before surrendering their properties! Chairman of the Landholders Committee was Alexander Patterson, the very same Patterson who had marched with Sheriff John Jennings of Easton, bringing "that terrible four-pounder" cannon to Wyoming in 1769.

The Committee of Connecticut Settlers responded the same day, "We do not think the lawful defense of what we esteem to be our own, can with any justice be termed a disaffection to Government...And although we mean to pay due obedience to the constitutional laws of Pennsylvania, we do not mean to become abject slaves, as the Committee of Landholders suggest in their address to your honors."

Chapter 18

"I looked back with this thought: shall I abandon Wyoming forever? The reply was, No! Oh, No!..."

The Second Pennamite War began with a serious of small incidents guaranteed to infuriate and insult the Connecticut-turned-Pennsylvania settlers.

The Board of Commissioners, appointed by the Pennsylvania Legislature to take testimony on individual land ownership claims of the former Connecticut citizens arrived at Wilkes-Barre in April, 1783.

Soon after, the commissioners created new political subdivisions, dividing Wyoming into three townships: Shawanese, Stoke, and Wyoming, all of Northumberland County.

By October of 1783, Alexander Patterson, of the Pennsylvania Landholders Committee, had been appointed a magistrate for the area and taken up residence in Stoke township, renaming Wyoming/Wilkes-Barre "Londonderry."

From Londonderry, Patterson lost no time arresting that "flagrant offender" Zebulon Butler and referring to all of the original Connecticut settlers as "banditti."

Conditions under Patterson became so intolerable the inhabitants of Wyoming were finally able to prevail on the Pennsylvania General Assembly to appoint another commission. This was done and on December 9, 1783, Jacob Stroud, Jacob Arndt, Jonas Hartzel, and Robert Brown, all of Northampton County (and from Stroudsburg) were instructed to go to Wyoming/Wilkes-Barre to inquire into charges made by inhabitants of Wyoming.

Afterwards, Robert Brown reported to the Legislature that "he had visited Wyoming as one of the committee upon the subject and had heard all the evidence upon both sides. The wrongs and sufferings of the people of Wyoming, he was constrained to declare, were intolerable. If there ever on earth was a people deserving redress, it was those people."

But the matter was tabled; the General Assembly session ended, time passed, and ice jams and floods along the Susquehanna wreaked havoc with people's lives and possessions - all of these events inured to the benefit of Magistrate Patterson. In April of 1784, Patterson directed soldiers to remove Connecticut survey markers, and on May 13 and 14 ousted 150 families with Connecticut title to their land, setting fire to their homes.

Patterson's men would not allow the refugees to travel by boat on the Susquehanna River, nor to take the road southeast to Stroudsburg, they directed the dispossessed to "take the Lackawaxen Road, as leading most directly to Connecticut."

"*But this way consisted of 60 miles of wilderness, with scarce a house; the roads wholly neglected during the war.*"

(How quickly Pike and Wayne Counties had become an undesirable "wilderness," now that the area once known as The Delaware Company's lands had failed to yield up tell-tale signs indicating the presence of "the black stones that burn.")

About the trek east, Elisha Harding wrote "It was a solemn scene: parents, their children crying for hunger - aged men on crutches -

all urged forward by an armed force at our heels. The first night we encamped at Capouse [Scranton]; the second, at Cobb's [today's Mt. Cobb exit on I-84]; the third at Little Meadows, so called [Hamlin]. Cold, hungry and drenched with rain, the poor women and children suffering much. The fourth night, at Lackawack [today's Paupack]; fifth, at Blooming Grove; sixth, at Shehola; on the seventh, arrived at the Delaware, where the people dispersed, some going up and some down the river. I kept on east, and when I got to the top of Shongum mountain [New York],

"I looked back with this thought: shall I abandon Wyoming forever? The reply was, No! Oh, No!

"There lie your murdered brothers and friends. Dear to me art thou, though a land of affliction.

"Every way looked gloomy, except towards Wyoming.

"Poor, ragged and distressed as I was, I had youth, health, and felt that my heart was whole.

"*So I turned back to defend or die.*"

Those who did return were forced to live on the cliffs of the eastern mountains, as their seized properties were in the possession of strangers. They selected a cave for headquarters, dubbing it "Fort Lillo-pe."

"In no part of the Union [the United States] were the sympathies of the people more generously aroused then among the just and good people of Pennsylvania." The troops that had ousted the settlers were ordered dismissed and all of the ousted Connecticut people were invited to return.

But Magistrate Patterson, for the Landholders Committee, formed his own military force, and on July 20, 1784, opened fire on a group going to the fields near Kingston.

Following Patterson's action, John Franklin, an early settler from Connecticut, mustered about 60 men. Starting at Kingston, Franklin and his men first marched down the west bank of the Susquehanna River as far south as Nanticoke, then, crossing the river, marched northward on the east bank "dispossessing every Pennsylvania family."

Patterson retaliated, and open warfare raged. The Sheriff of Northumberland County rushed to Wilkes-Barre to restore peace; Patterson called for reinforcements from among friends in New Jersey; the Pennsylvania Supreme Executive Council called in troops; and a small detachment of Connecticut-now-Pennsylvanians, hearing rumors of approaching troops and believing them to be for Patterson, rushed to Locust Hill (near Stoddartsville where Route 115 crosses the Lehigh River). The skirmish at Locust Hill left one killed and several wounded.

When the Connecticut fighters were brought into court at Easton (the Locust Hill battle had taken place in Northampton County), the jury returned a finding of "ignoramus," mistake in facts. "The jury considered the settlers as fighting in self-defense; and such was the popular opinion throughout the State." Likewise, juries at Northumberland refused to return indictments against those Connecticut settlers jailed at Sunbury.

Many attempts were made to stop the fighting. The Pennsylvania Legislature, under a date of September 15, 1784, ordered the settlers to be restored to their possessions, but before that actually happened, General Assembly defied the Council of Censors, refusing to surrender documents, and the militia of Bucks, Berks, and Northampton Counties "generally declined obedience to orders," and would not march to Wyoming.

Sporadic fighting continued until Saturday, November 27 when it was recorded that "the Pennymites <sic> evacuated the fort [Forty Fort] at 11 p.m." and on the following Tuesday, November 30, "the Yankees destroyed the fort."

The year 1784 ended with the Yankee Pennsylvanians the sole inhabitants of the portion of northeastern Pennsylvania that counted, the area where "the black stones that burn" were known to exist. Elsewhere, popular opinion supported the Yankees in their desperate struggle to hold onto the land that they had purchased with their own money and many of their lives.

Chapter 19

Who would own the "black stones that burn"?

The hunt for anthracite coal, "the black stones that burn," was on in earnest. Gabriel Plattes authored a 37-page pamphlet, published in 1784 in Philadelphia, titled *A Discovery of Subterranean Treasure: Containing Useful Explorations Concerning All Manner of Mines and Minerals from the Gold to the Coal.* According to Plattes, there were five signs indicating the presence of coal: the tops of barren ridges of rocky barren mountains, the presence of black slaty rocks, the color of grass, the smell of vapors, and spring water with a black sediment.

According to mining expert Samuel Hermelin, a place with all those features had been found in 1783 on top of Broad Mountain at Beaver Meadows (south of today's Hazleton) and loose pieces of coal had been found lying on the ground.

So maybe, just maybe, the coal of the Wyoming Valley existed elsewhere.

When the 2nd Pennamite War ended as a victory for the Yankee Pennsylvanians, the Pennsylvania Landholders backed off. Public opinion was definitely in support of the Yankees holding on to the farms and tracts of land they had purchased from the Connecticut-based land companies.

But the embattled Yankee Pennsylvanians were greatly distraught. John Franklin, who had ousted Pennsylvanians after Patterson's expulsion of the Yankees, made three trips to Hartford, Connecticut over a five month span.

When he returned to Wyoming from his third trip, Franklin reported that on July 13, 1785, the Susquehanna Company, as owners/developers of the tract of land along the Susquehanna River purchased from the Indians when Connecticut was the governing political body, had voted to support all claims and Rights-of-Soil ownership; and was actively encouraging future settlements within the area.

By the following April 8, 1786, Franklin noted that Ethan Allen, Revolutionary War hero and champion of statehood for Vermont, visited Wyoming and "proposed to settle amongst us and to bring on with him a number of his Green Mountain Boys, and assist us in supporting and defending our rights against the Pennsylvania Claimants."

A large tract of Susquehanna Company land was given to Ethan Allen for that purpose.

Much like the private communities of today, property owners within the Susquehanna Company development set up their own governing council and promulgated their own laws, suitable to the community.

It became a race against time whether or not Pennsylvania could prevent dismemberment. The inhabitants of Susquehanna Company territory were ready to establish their own state.

"The first grand healing measure adopted on the part of" Pennsylvania was to pass an Act on September 25, 1786, erecting the northern portion of Northumberland County into a new county called Luzerne. The new county had once been the western half of Connecticut's Westmoreland County.

(What was once the eastern portion of old Westmoreland, including the settlements at Lacawa (Paupack) and along the Delaware River, remained in Pennsylvania's Northampton County, as far outposts of the civilized county seat at Easton.)

The name Luzerne was selected perhaps because it was synonymous with the grace and charm of France's Minister Plenipotentiary to the United States, the Chevalier de la Luzerne, from November 17, 1779 to 1784. The Chevalier had hosted a party on July 16, 1782 at his Philadelphia residence that was "never equalled."

To set the new county in motion, Timothy Pickering was sent to Wyoming to join Zebulon Butler and John Franklin. Pickering was a native of Massachusetts who had served as Quartermaster-General under Washington during the War of 1776 and afterwards settled in Philadelphia. He was looked upon as one of the "most eminent men in the Union." Zebulon Butler and Franklin, both early settlers at Wyoming, had been in the Wyoming Massacre. The three men, by the Pennsylvania Act of December 27, 1786, were named "to organize the new county, and introduce the laws of the State among the Wyoming people."

Would Pennsylvania's "band-aid" do the trick?

Franklin, however, was not interested in seeing a separate county for Wyoming. He was too devoted to the concept of a separate state. Soon, for the first time, the Yankees would divide.

Zebulon Butler and Matthias Hollenback gave their support to Pickering and Luzerne County; Franklin, elected to represent Luzerne in General Assembly (but would not serve), urged fidelity to Connecticut title and Susquehanna Company. He was backed by the Jenkins and Spaldings.

The biggest issue was the Confirming Law. Passed March 28, 1787 by Pennsylvania Assembly, the Act stated that rights-of-soil of all Connecticut settlers prior to the Decree of Trenton "are hereby confirmed to them and their heirs and assigns." A commission would review the claims.

Franklin believed titles should be confirmed before there was any thought of taking an oath swearing allegiance to Pennsylvania.

A public debate held at Forty Fort ended in a melee when Hollenback, a newly elected Luzerne County judge, flicked his riding whip at Franklin's head as Franklin spoke.

Pickering, gravely concerned at the turn of events, shortly afterwards arranged secretly to have Franklin arrested and taken to Philadelphia, to be jailed there.

"All Wyoming was in commotion from Nescopeck to the [New York] State line, on hearing of the abduction of Franklin, and the part Colonel Pickering had acted in the scene."

Eventually, Pickering was abducted by Franklin supporters and carried up the Susquehanna River, north of Tunkhannock. Three weeks later, Pickering was released. Franklin, on the other hand, would remain in jail at Philadelphia for a year, until November 1788.

In the end result, however, Franklin was proven to have been right when he argued for Yankee right-of-soil approval before swearing loyalty to Pennsylvania and Luzerne County.

By the terms of the Confirming Law, the Connecticut settlers, who were scattered all over Northeastern Pennsylvania, had just eight months in which to gather proof of their original land ownership and present it to the Commission sitting at Wilkes-Barre. But "the disturbances of the country already detailed, occasioned the withdrawal of the commissioners appointed to carry it into effect."

On April 1, 1790, the Confirming Law was repealed by another act of Pennsylvania Assembly. (Five years later, in a land eviction lawsuit, a United States Circuit Court Judge ruled the Confirming Law unconstitutional.)

Pickering later wrote "By this repeal [of the Confirming Act], the courts of law were opened to the Pennsylvania claimants."

Who would own "the black stones that burn"?

Chapter 20

Yes! There were "black stones that burn" beyond the Connecticut claim.

There was much land speculation during the 1780s. The war was over, the union of states was now the United States of America. Everywhere throughout the 13 states an air of optimism and vitality prevailed. Men believed great fortunes could be made supplying the needs and wants of the new country. Philadelphia, capital of both the nation and the "keystone" state, was particularily exhilarating during the 1780s.

Yes, indeed, Philadelphia was an exhilarating place to be, a mecca for men of ideas and for men seeking financial backing for a project.

And more people were learning that northeastern Pennsylvania was the only place where the unusual "black stones that burn" - anthracite coal - could be found.

John Nicholson, born in Wales, spent part of his youth in Carlisle, Pennsylvania before coming to Philadelphia. He displayed great ability in financial matters and was appointed comptroller-general of Pennsylvania in 1782. He served in that capacity, as well as in additional positions of trust and influence, until 1794.

To Nicholson and his coterie of about 100 men, all active in Pennsylvania affairs, must go the credit for bringing about a satisfactory resolution of the Wyoming struggle.

The men were all patriots; on many of them hung the success of the American Revolution. They were entrepreneurs, too, mainly as land speculators. Counting on an ever increasing population, they envisioned themselves as the suppliers who would meet both the domestic and the military needs of the new nation, with a little left over for export. And along the way, they expected to accumulate great wealth.

One of Nicholson's business associates, later his friend, was Robert Morris, a signer of the Declaration of Independence. Morris had supervised the finances of the War of 1776 and established the Bank of North America in 1781.

Nicholson and some of his closest friends and business partners had been able to secure vast tracts of land in western Pennsylvania (today's Beaver, Allegheny, Butler, Armstrong, Lawrence Counties) starting in 1783.

The area was part of the Pennsylvania lands that had been confiscated from the William Penn heirs in 1779. The state released it to help pay off the more than $6 million liability that Pennsylvania had assumed from the federal government as its share of the national war debt.

This was also an area where pioneers had been using soft (bituminous) coal to heat their cabins because of "the abundance of surface coals."

Building on early purchases, working with Daniel Brodhead, Pennsylvania Surveyor-General (the same Brodhead who had marched up the Allegheny River when Sullivan swept the Indians from Elmira, New York, and who eventually settled in Milford, Pike County), Nicholson soon gained a reputation for great wealth and a willingness to back major projects. Strangers as well as friends constantly besieged him for financial support. The gregarious Nicholson rarely refused.

Nicholson and his fellow land speculators were often accused of selling inferior lots to settlers and holding back the most desirable parcels for themselves. But whether or not Nicholson's sole interest in land deals was just to open up "wild lands" for settlement, as he openly avowed, it must be observed that all of his landholdings were in areas where the presence of great deposits of minerals were suspected.

Connecticut claims in northeastern Pennsylvania were a problem to Nicholson. Buying Pennsylvania title where he could in the 1780s, Nicholson "wrote to one of his land agents that the Pennsylvania claims should take precedence over the intruders [Connecticut Yankees]." To his wife, Nicholson advised her "not to worry as he was well armed against the Yankees." The developer claimed "a monopoly on the town lots and on the surrounding mountains in Wilkes-Barre."

Who would eventually own "the black stones that burn"?

The Society for Promoting the Improvement of Roads and Inland Navigation was organized in March, 1789 at Philadelphia by John Nicholson, Robert Morris, William Bingham, John Adlum, James Wilson, Henry Drinker, and Thomas Mifflin, all major land developers.

Mifflin would become governor of Pennsylvania the following year, and James Wilson, like Morris a signer of the Declaration of Independence, would be appointed an associate justice of the U.S. Supreme Court before 1789 ended. Drinker was in the process of sending his partner Samuel Preston to Stockport (Equinunk, Wayne County) to oversee the development of "Harmony" (now in Susquehanna County).

One of the major activities of the Society was to apply pressure to Pennsylvania Legislators to appropriate funds for improving the state's transportation system - a "hidden" subsidy for developing their lands.

The course was set then which would slowly, inexorably remove the Yankee "banditti" from their Pennsylvania landholdings in a quiet way, without another Pennamite war.

That is, if it had not been for Philip Ginter in 1791.

Ginter, operator of a mill, was roaming the countryside seeking a new millstone. On Sharp Mountain, near today's Summit Hill (about 15 miles west of today's Jim Thorpe in Carbon County), he noticed an unusual black stone. Ginter managed to break off a small chunk.

The miller traveled 20 miles to Jacob Weiss, "the most prominent settler in the area," who lived at Weissport (near Lehighton) to get his opinion on the black stone.

Weiss in turn traveled to Charles Cist, his brother-in-law in Philadelphia, with the specimen of anthracite coal.

The Lehigh Coal Mine Company was organized in Philadelphia on February 13, 1792, including as shareholders Weiss, Cist, John Nicholson, and Robert Morris. (Ginter would not accept stock in the company.)

Yes! There were "black stones that burn" beyond the Connecticut claim.

Summit Hill and Weiss-port are all on approximately the same degree of north latitude, 40°50', thereby placing Ginter's coal discovery about ten minutes of a degree past Connecticut's famous southern boundary line - Latitude 41° North. (Stroudsburg is at approximately Latitude 40°59' North.)

Let the Yankees have their land!

Chapter 21

At Last! The Yankees had their land!

While the 1790s was a period of great activity, the decade was really a marker for the end of man's way of life as he had known it from the beginning of his existence on earth.

And all because James Watt, at Glasgow, Scotland, had perfected the steam engine so that it was commercially viable. With a partner, Matthew Boulton, Watt had manufactured about 500 of the engines by the year 1800.

The importance of the steam engine cannot be overemphasized.

Why?

Because it freed man from the earth.

Man was no longer tied to the earth.

When the Industrial Revolution began, man's only outside source of energy and power was the water wheel. Fast flowing streams, such as the Merrimack River in Massachussetts, could turn a water wheel, which, when hooked up to other wheels, gears, shafts, and axles, provided the energy to manufacture a product better and faster than several men working together.

But with only water as a source of power, factories and machines could never be very far from a river bank. Industrial communities could only develop clustered along a special waterway, as at Lowell, Massachussetts, where the Merrimack flowed its fastest.

Whereas the steam engine provided energy wherever man wanted to go.

Eventually, movable steam engines (on railroads) would transport fuel to operate stationary steam engines; and sprawling cities like Los Angeles, with almost no year-round fast moving streams, could have manufacturing everywhere.

Boulton-Watt engines were first used in England to hoist excess water as well as ore from mines. Their second application was to operate the bellows in an iron blast furnace. From the beginning, bituminous (soft) coal was used for the steam engine, although wood would also work. This was because soft coal was plentiful in the British islands.

At the same time that the 1790s saw a growing need for the steam engine in England, the 1790s was also bringing discovery of additional anthracite (hard coal) deposits in the United States.

What would happen if Great Britain's version of the steam engine and America's anthracite coal ever got together?

Another major anthracite coal field was apparently discovered by Necho Allen near Pottsville in 1790, a year before Ginter brought his sample to Jacob Weiss. Allen, however, kept his find secret until he had secured enough land to do his own mining.

Ginter's coal deposits near Jim Thorpe, and Allen's near Pottsville, were in the watersheds of the upper reaches of the Lehigh and Schuylkill Rivers respectively. These rivers linked coal lands more directly and easily to Philadelphia than did the Susquehanna River. So Philadelphians concentrated on developing the coal fields closer to them, especially as the land titles were not so involved as with the Yankees at Wyoming.

For John Nicholson and Robert Morris, the indefatigable land speculators of Philadelphia, coal near Pottsville and also near Jim Thorpe could mean coal elsewhere. Apparently applying the 1784 instructions of author Gabriel Plattes in his 37-page pamphlet - to look for coal, seek bare ridges on barren mountain tops, black slaty rocks, etc. - and extending an imaginary line northeastward, a portion of present day Pike County caught their eye.

In 1794, the two real estate magnates acquired over 60,000 acres in what are now the five Pike County townships of Blooming Grove, Dingman, Greene, Palmyra, and Porter. (Even today some deeds from that area carry the reservation clause that minerals are "excepted out.")

And if it hadn't been for a trick of "Mother Nature," who knows - maybe Nicholson and Morris would have guessed correctly. The science of geology was in its infancy in the 1790s and little was known about rock "folds," and "faults."

The anthracite beds of northeastern Pennsylvania were all formed at the same time, over 250 million years ago. Subsequent upheavals that created the mountains (folds) also caused rock movement (faults). Who knows what shape the coal beds may have taken, or even formed one large bed, had it not been for the eruptions of earth's crust, especially rock movement?

(A dramatic example of rock movement can be seen today on Interstate-81 at mile 138 in Schuylkill County. A deep road cut in Spring Mountain when building the highway exposed one large, high-angle thrust fault and two low-angle thrust faults. "The basic structure is a syncline in which the rock layers are broken by faults." The rock movement, in this instance, is a little over 50 feet.)

As it turned out, Ginter's discovery near the community of Jim Thorpe/Mauch Chunk was at the eastern edge of one coal field. The field did not extend any further eastward, so the Nicholson/Morris lands would not have the kind of value the speculators had hoped. Nicholson died in 1800, and Morris in 1806, without developing their real estate in Pike County, probably never knowing that it was not coal rich.

Elsewhere, Nicholson, at first with Morris, then later by himself, owned over 300,000 acres along the Susquehanna River about 50 miles north of Wilkes-Barre. The two men must have sensed there was also little coal there, because in 1793 they offered it to French refugees escaping their own revolution. The community is called Azilum, partly in Bradford County and partly in Wyoming County, at that time the northern end of Luzerne County.

So with the coal discoveries at Pottsville (then Berks County, now Schuylkill County), and at Jim Thorpe/Mauch Chunk (the western edge of Northampton County, now Carbon County), the Yankees and Wyoming looked less enticing. As a result, the Pennsylvania government took a softer and softer approach to the Right-of-Soil question at Wyoming.

Finally, the Compromising Law was passed on April 4, 1799. Sometimes called the "Compensation Law," it, almost word for word, conferred the same status that the repealed Confirming Law of March 28, 1787 had on the Connecticut settlers. Namely, that Connecticut claimants of the 17 townships erected within Susquehanna Company lands who had been actual settlers before the Decree of Trenton (December 30, 1782) held valid title to their properties. A supplement passed seven days later on April 11, aimed at stopping any further Connecticut settlements. A fine of $1,000 and imprisonment would be levied on anyone who could not produce a Pennsylvania title.

Under the terms of the Compromising Law, commissioners were to review each Connecticut claim and issue a "Certificate". At the same time, all Pennsylvania claimants, where they conflicted with Connecticut title, were to receive monetary compensation for vacating the contested land.

Nicholson, who had held Pennsylvania title to the Azilum tract, had also bought the Connecticut title in 1795. With the 1799 Compensation Law in effect, he was eligible for nothing.

At Last! The Yankees had their land!

Chapter 22

Now it was time to be about the business of turning the new United States of America into a world power.

So, in the end, it would be the entrepreneur who would be entrusted with the responsibility of turning his new country into a world power, thus making the Industrial Revolution in America the product of individual men - their creative minds and their charismatic personalities.

For Pennsylvania landowners whose properties did not conflict with original Connecticut claims, the Commonwealth of Pennsylvania issued "original patents." All unclaimed land remained with the state, which issued a patent whenever a tract was sold.

When it became apparent that the land question would soon be resolved, prospective settlers started moving into northeastern Pennsylvania at an increasing rate. By 1794 there were enough inhabitants in what had once been the eastern half of

Connecticut's Westmoreland County to demand a county of their own. The residents insisted that the trip to the county seat at Easton was too much of a hardship. Pennsylvania Assembly responded by erecting the northern portion of Northampton County into Wayne County on March 21, 1798, naming it for General Anthony Wayne of Revolutionary war fame. Milford was designated as the county seat.

With the establishment of Luzerne and Wayne Counties, Connecticut's claim passed into an almost forgotten memory. Except for that indefatigable upholder of Locke's political theories, John Franklin, survivor of the Wyoming Massacre. Franklin, elected sheriff of Luzerne County and several times to Pennsylvania Assembly, retired from the Wilkes-Barre area, moving north to Tioga Point, near today's Athens. He continued a strong advocate of all things Connecticut.

Finally, "For the purpose of legislating Colonel John Franklin out of the Assembly, to which the people of Luzerne persisted in sending him, and where his earnest and persevering advocacy of the claims of the Connecticut settlers rendered him exceedingly obnoxious to those holding Pennsylvania titles in his district, and to the Pennsylvania Landholders' Association, which exerted great influence in the Legislature, an act was passed April 2, 1804, setting off so much of Luzerne"..."including the district in which Col. Franklin lived" and attaching the area to Lycoming County.

(This portion of Lycoming County then became a separate county on February 21, 1810, designated Bradford County in 1812. The county seat is Towanda. Lycoming County had been erected on April 13, 1795, and at that time all of the northwestern portion of Pennsylvania to the Allegheny River was included in Lycoming.)

But Franklin could not be kept down. "After several efforts one proved successful, and behold the great Yankee leader was chosen a representative from Lycoming County, appeared in triumph at Lancaster, and took his seat December 1805."

Franklin and other early Wyoming settlers throughout northeastern Pennsylvania continued to rise to positions of honor and trust in administering the laws of Pennsylvania, until finally the Connecticut flame that had given the early settlers their strength to endure the greatest of sufferings and privations became a glowing ember of remembrance.

By the turn of the century (early 1800s), good and valid Pennsylvania title could be established for all property owners in northeastern Pennsylvania.

Now it was time to be about the business of turning the new United States of America into a world power.

After the Revolutionary War ended, New York found that it had a new wealth: white gold - salt.

Today's Syracuse, New York, is at the very heart of the Iroquois Confederation. In the 1600s and 1700s, visitors to the Onondaga Indians' council fire had always been cautioned to stay away from the water of Lake Onondaga because it was "bad." But after the War of 1776, with many Indians relocated in Canada and the Onondagas restricted to a reservation south of Syracuse, the white settler at last had a chance to test the waters of Lake Onondaga for himself - to find a vast salt deposit.

So, the white gold that was salt joined the green gold that was unlimited forests of timber, the black diamond that was coal, and the mineral with the tell-tale reddish stain that has given its name to a prehistoric age - iron. All were found in great quantity but far from civilization and commercial markets.

To be viable, the federal government and/or state governments, or private enterprise, via entrepreneurs, would have to find a way to make these natural resources accessible.

When all the states in the Union tallied their assets after the Revolution, Pennsylvania found it had an abundant supply of everything except salt. It was truly a "keystone state," not only because of its geographic location among the 13 original states but also because of its key natural resources.

And by 1800, Pennsylvania had solved its first problem - who would own the various natural resources within its borders? The next problem was transportation - once a resource was harvested, how could it be brought to the consumer? The federal government did not play an active role in resolving the issue of transportation. It finally came to rest with private enterprise, the entrepreneur, who in turn depended on state governments as much as possible.This might be described as an unexpected turn of events, because many of the first men attempting to develop Pennsylvania's resources had been active in establishing the federal government.

Robert Morris, the Revolution's financial adviser, chose to align himself with John Nicholson, Pennsylvania's comptroller-general, acting as private land developers rather than advocating federal involvement. This philosophy was emphasized when Morris, Nicholson, and a New Englander, James Greenleaf, acquired 42% of all building lots in Washington, D.C. when the nation's capital was first developing in the 1790s.

Letting private enterprise manage the country's assets was an interesting arrangement, probably indicative of the very intense popular support for states' rights. It was definitely touted as a means of keeping taxes at a minimum. But more possibly, it showed that individuals, once they had thrown off the royal yoke for democratic equality, and were free from subjugation to aristocracy's hereditary land rights and its protective paternalism, had in their place substituted self-reliance, making its outward sign - success - a virtue.

Francis A. de La Rochefoucault-Liancourt, traveling in America from 1795 to 1797, noticed this when he wrote after his trip, "In America where, more than any other country in the world, a desire for wealth is the prevailing passion, there are few schemes which are not made the means of extensive speculation..."

So, in the end, it would be the entrepreneur who would be entrusted with the responsibility of turning his new country into a world power, thus making the Industrial Revolution in America the product of individual men - their creative minds and their charismatic personalities.

Chapter 23

In many ways, the Moravians were like the Quakers.

There was a universal air of excitement in the early days of the American Industrial Revolution. Everyone recognized that it was possible for every person, for any person, to invent, to advance mankind. Opportunity was open to all

As long as the social contract among people had been administered according to the laws of feudalism and aristocracy, birth, brawn and might were the determining factors. But once man was no longer tied to the land, intelligence and creative ability became dominant under an assortment of social orders.

Disillusionment and disappointment with the first promise of the Industrial Revolution would not come until the last quarter of the nineteenth century, when the gap between successful inventors and promoters and the rest of the laboring population became very, very noticeable, with the admired creative genius turning into the despised robber baron.

In the early days of the Industrial Revolution, the general perception was that it was possible for anyone to acquire great wealth. The Industrial Revolution would come to an end with the recognition that great wealth would come to only a few. In the beginning, religion would be the catalyst; in the end, religion would provide the excuse.

When the Spaniards came to the new world following Columbus' discovery in 1492, the gold and silver deposits in Central and South America were so spectacular that no land was considered worthwhile unless it could match these finds.

Spanish explorers came into North America, criss-crossing the southern portion several times, but no gold or silver was ever found to equal or surpass what was already known to exist in Central and South America. North America became expendable, then - a throw-away land.

And after Captain John Smith reported on his travels in North America to the English king and people, that the only wealth of the land lay in its fish, fur animal pelts, and virgin timberland, what an easy thing it was to give the land away to troublesome religious zealots. So the Pilgrims came first, to Plymouth Rock, Massachusetts in 1620, with their belief in freedom of religious conscience, and civil authority never abdicating to the church - to create the *Mayflower Compact.* And another religious dreamer, William Penn, with his ideals of peace, was given the territory between Lord Baltimore's and the once Dutch colony of New York: Penn's woods - Pennsylvania - in 1681.

Penn used his gift from the English king, Charles II, to put into practice the beliefs of his new religious fervor - that all persons may perceive the word of God in their soul. Penn did not start or found a religion or denomination. What he did was to offer safe haven to all religious creeds.

He, and many others who had become followers of George Fox, an English lay preacher, were eventually known as the Religious Society of Friends, sometimes called Quakers. As Fox had started preaching in England in 1647, there were many Quakers in the

American colonies long before Penn's arrival. These Quakers, mostly living in the northern, New England, colonies were often despised and discriminated against by Puritans (Massachusetts Bay Colonists), and the New York Dutch.

(In 1666 a member of the Chelmsford Puritan church, at Lowell, Massachusetts, "was publicly admonished by a vote of the church members after they received reports that he had been 'at Rhode Island among the Quakers.'")

For Quakers in Pennsylvania, however, the environment was entirely different. For one, Quakers were in the majority (approximately 7,000 by 1684). The other, most important difference was that when Penn began his career as a land developer, he wrote *Frame of Government*, guaranteeing freedom of worship, protection of property, and trial by jury to all who would settle in his "Woods."

Many English Quakers took Penn up on his promise of freedom and came directly to Pennsylvania, so their only experience of life in the New World was as a respected Quaker. But New England Quakers had been Friends long before Penn ever arrived at Philadelphia. They had had to learn to live as a persecuted minority. While they were Quakers, too, they were worlds apart from Penn's Quakers.

Penn's *Frame of Government* attracted many religious groups, some similar to the English Quakers. One group, originally from Czechoslovakia, came to Pennsylvania by way of Georgia. In 1741, "a nucleus of extraordinarily committed Moravian settlers came together on the banks of the Lehigh River to build a new society devoted to religious missions."

In many ways, the Moravians were like the Quakers.

They were pacifists, abstaining from all forms of combat or war, and their leader Count Nicholas Ludwig von Zinzendorf started the Moravians on "a far reaching missionary program to bring Christianity to the 'heathen.'" It had been from their settlement on the Lehigh at Bethlehem that the Christian Indian communities of Gnadenhuetten (near Jim Thorpe in Carbon County) and Friedenshütten (near Wyalusing on the Susquehanna River in Bradford County) had been started.

And in the wilderness that was northeastern Pennsylvania, there was little difference to the convert between the Quaker and the Moravian missionary.

As was the case with a small group of Minsi Indians living at Wyalusing (later Friedenshütten) in the 1760s. Dissatisfied with their chieftain Papunhank's teaching of Christianity, learned from visits to Philadelphia, as it contrasted with "his own bad life," they decided to invite "a white teacher to settle among them."

"In their councils, however, they were divided in opinion, one party in favor of the Quakers, and the other of the Moravians, and so equal was the strength of the two parties that neither was disposed to yield to the other. Their differences were compromised by agreeing to accept the first teacher who came.."

The Quaker evangelist John Woolman and the celebrated Moravian apostle David Zeisberger both left for Wyalusing. The two missionaries dined together south of Wilkes-Barre, but it was Zeisberger who arrived at Wyalusing on May 23, 1763, two days ahead of Woolman.

Zeisberger "was received as the divinely sent messenger, to teach them the great words of the Christian religion." Woolman, for his part, was kindly received but stayed just five days "to assist in inaugurating the good work" at what was now Friedenshütten.

As for the Quakers, their missionary work was universally recognized by the Indians, as in the case of Isaac Tripp, one of the forty Connecticut settlers who returned to Wyoming in 1769:

"As Tripp was a man of more than ordinary efficiency and prominence in the colony, the Indians were often asked by the British why he was not slain. The unvarying answer was that Tripp was a good man. He was a Quaker in his religious notions, and in all intercourse with the Indians his manner had been so kind and conciliatory that when he fell into their hands as a prisoner the year previous at Capouse [now Scranton], they dismissed him unharmed."

Chapter 24

It would be from the ranks of these two religious groups - the Society of Friends and the Moravian Church - that the true entrepreneurs of Pennsylvania's industrial might would come.

From the Moravians would come Jacob Cist, and later William Henry bringing the Scrantons to the Pennsylvania city to bear their name.

Jacob Cist was the son of Charles Cist and the nephew of Jacob Weiss, all Moravians. In fact, Jacob Cist had been sent from Philadelphia to Nazareth, Pennsylvania, to be educated at the Moravian school there.

It was Jacob Cist who would inherit shares in the Lehigh Coal Mine Company started in 1792 near Mauch Chunk (now Jim Thorpe) by his father and uncle, with the aid of "reputable Quaker merchant" John Nicholson and Declaration of Independence signer Robert Morris. The Lehigh Coal Mine Company opened up the eastern end of the southern anthracite coal field.

And from the Quaker village Mount Holly, New Jersey (halfway between Philadelphia and Trenton), Josiah White would come to Philadelphia and continue to Mauch Chunk (*Today's* Jim Thorpe). As would the Quaker philanthropist Stephen Girard, who from Philadelphia would control a goodly portion of the coal fields north of Pottsville.

And it would be a few Philadelphia Quaker extended families, such as the Wetherills and the Leas, who would control all of the western portion of the southern coal field around Pottsville.

(The anthracite middle coal fields - generally Hazleton to Shamokin - while suspected, would not be opened until the 1830s.)

The northern anthracite coal field, however, was generally in possession of the Yankee Pennsylvanians who, while Christians, seemed to be free from religious domination. Many were of Puritan background; quite a few were from the current generation of Pilgrims. Were there no Quakers? Would there be no Moravians?

It would be the Moravian Jacob Cist who in 1807 would marry Sarah Hollenback, daughter of Matthias Hollenback, a Wyoming Massacre survivor and shipper of "stone coal" to Carlisle. Cist would come to live at Wilkes-Barre and would claim a tie to both the southern and northern anthracite coal fields.

And in 1785, Jesse Fell, a Quaker from Bucks County (immediately north of Philadelphia), would come to Wilkes-Barre to keep a store and tavern.

And leading the way from Rhode Island, Connecticut and New York, purchasers of original shares in the Susquehanna Company, were two New England Quaker families that had intermarried, Tripp and Slocum. They would eventually choose Providence Township (Scranton) on the banks of the Lackawanna River about 20 miles northeast of Wilkes-Barre, to make their settlement.

(Frances Slocum was a child of five when carried off by Indians on November 2, 1778. Nothing more was known about her until 1835, when Colonel G. W. Ewing, of Logansport, Indiana, wrote:

"There is now living near this place, among the Miami tribe of Indians, an aged white woman, who, a few days ago, told me that she was taken away from her father's house, on the Susquehanna River, when she was very young. She says her father's name was Slocum; that he was a Quaker, and wore a large-brimmed hat; that he lived about a half a mile from a town where there was a fort [Wilkes-Barre].")

And in 1812, Philadelphia Quakers Maurice Wurts and his brother William would be "patiently exploring the mountain wilderness of Pennsylvania in a practical, painstaking search for the true philosopher's stone. Following the Lehigh from Mauch Chunk far up into the forest to its very headwaters in the Pocono marsh, only to be baffled in their search for a new and unclaimed coal field, they struck at length into the almost equally wild and sparsely settled valley in which the Lackawanna runs, and without any previous knowledge of the region, traversed it for months, seeking everywhere for traces of the buried treasure."

(Years later, Paul Preston would recall hearing his Quaker father, Samuel Preston of Stockport on the Delaware River in Wayne County - while in Philadelphia in 1814 - tell Maurice Wurts, "Maurice, thee must hold on to that lot on the Lackawanna that you took for a debt of David Nobles; it will be very valuable some day, as it has stone coal on it and under it.")

Yes, there were Moravians and Quakers, who would come to Northeast Pennsylvania.

Some Quakers, like Samuel Preston, would work with his Philadelphia partner, Henry Drinker, to develop Wayne and Susquehanna Counties with the Stockport and Harmony settlements. And another Quaker from the Philadelphia area (Burlington, New Jersey, near Mount Holly), was Richard Smith, who, starting as early as 1769, would turn the area at the headwaters of the Susquehanna River (Cooperstown, New York) into a Quaker enclave, Quakers from there coming into northeastern Pennsylvania. And from Philadelphia, later Trenton, New Jersey, Quakers Samuel Meredith and brother-in-law George Clymer, a signer of the Declaration of Independence, would make their contribution to northeastern Pennsylvania with their landholdings across northern Pennsylvania and into southern New York. (Meredith eventually moved to Belmont, Mount Pleasant Township in Wayne County and died there in 1817.)

And it would be the Quakers from the Oblong, that area between New York and Connecticut, who, with the blessing of their Meetings, would come to northeastern Pennsylvania and begin a Meeting there, on the banks of the Lackawanna River in Providence Township, where beloved Indian Chief Capouse had once guided his people.

As soon as it became apparent that Pennsylvania would issue good title to land in the old Connecticut claim area, members of the Griffin family, Quakers from the Oblong, started moving into Luzerne County. Between 1794 and 1825, three Quaker families, Griffin, Slocum and Tripp, would own 10,200 acres of the 16,000 acres that made up the original Providence Township - the area that became Scranton.

The first Griffin to arrive in Providence was Stephen. An entry in the record book of the Stanford, New York, Monthly Meeting Society of Friends shows that he and his family were approved for leaving Stanford to go to the meeting, (although not named,) at "Providence, Luzerne County, Pennsylvania."

And at Stroudsburg, always friendly to the Connecticut settlers, Daniel Stroud, son of Jacob Stroud, joined the Society of Friends in 1804, giving up his profession as a lawyer.

So, yes, Quakers and Moravians did come to the northern anthracite coal field, and would accomplish wonderful things to assure a rich and powerful future for the United States.

But along the way, the inaugural meeting of the Quakers at Providence would quietly slip into oblivion, and the members of the Griffin family would eventually be found on the membership rolls of Meetings near Muncy, at the junction of the East and West Branches of the Susquehanna River, and then at Stroudsburg Monthly Meeting.

Chapter 25

The entrepreneurial spirit of 200 years ago made a two-prong attack on the problem worthy of the most advanced marketing techniques of today: generate an intense need within the consumer; and assure reliable, dependable supply to the consumer.

All those with a vested interest in northeastern Pennsylvania knew that a lot of hard work lay ahead of them if they wanted to realize a profit of any kind on their investment.

The entrepreneurial spirit of 200 years ago made a two-prong attack on the problem worthy of the most advanced marketing techniques of today: generate an intense need within the consumer; and assure reliable, dependable supply to the consumer.

The goods the entrepreneur had to work with in northeastern Pennsylvania were: land for development; timber; and anthracite coal.

For his part, Jacob Cist, with his inquiring, scientific mind, "checked on the inventory," focusing on the "black stones that burn."

Cist mapped the anthracite coal fields. This was a truly remarkable feat because, as can be seen by a comparison between the two maps, Cist, self-taught and with rudimentary tools, generally identified the fields, including the unexpected shift northward that had eluded Nicholson and Morris when they purchased Pike County lands.

Cist also worked on creating a need for his product.

When Cist married, he went into partnership with his father-in-law Matthias Hollenback, moving to Wilkes-Barre in 1811. Hollenback, originally backed by John Nicholson, had eventually secured title to over 48,500 acres of land in both Pennsylvania and New York. In addition to land holdings, Hollenback operated mercantile stores thoughout the territory similar to trader outposts. Cist worked with his father-in-law on all projects.

Besides mapping the anthracite coal region, Cist wrote, in 1815, *On the Importance and Necessity of Introducing Coal, Particularly the Species Known by the Name of Lehigh Coal into Immediate and General Use.* The pamphlet was to be printed and re-printed over and over again.

He gave instructions to iron masters on using anthracite in their forges, and for brewing and distilling; also how to build a nail-making furnace, a lime kiln, a potash furnace, and a glass furnace; he designed heating and cooking stoves. He identified the two kinds of anthracite making up the coal beds - based on the color of the residual ash, white or red, the coal was either almost pure carbon or carbon with iron impurities.

In addition to his scientific work, Cist turned architect and drew the plan for the community that would become known first as Mauch Chunk and later Jim Thorpe, on the Lehigh River.

Talk about a salesman!

Cist conducted his scientific investigations into coal with the ultimate aim of attracting investors into Northeastern Pennsylvania and to have on hand reasons why anthracite coal should be used.

The first commercial use of the "black stones that burn" was with iron mongers and blacksmiths. But sometimes it was difficult selling the product. Bituminous, or soft, coal had been available from Virginia since before the War of 1776. And iron furnace plantations were traditionally located on vast tracts of land, providing the on-the-spot wood for fuel. Both these fuels could be obtained for a much lower price than the cost of hauling wagon loads of anthracite to a commercial center.

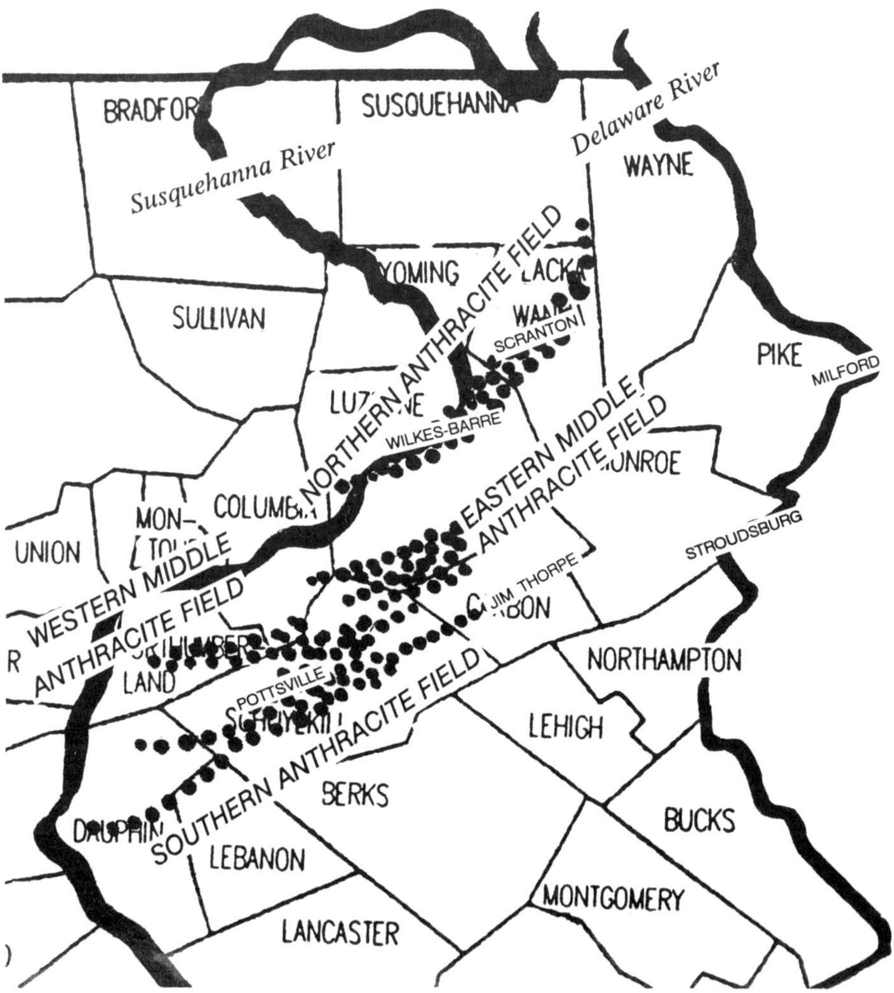

The War of 1812 changed all that, giving the anthracite promoters their opening wedge. Because the war was between the United States and Great Britain, all imports of British soft coal ceased. This put Virginia bituminous into short supply compared to the demand. Anthracite promoters lost no time in rushing in to fill the gap.

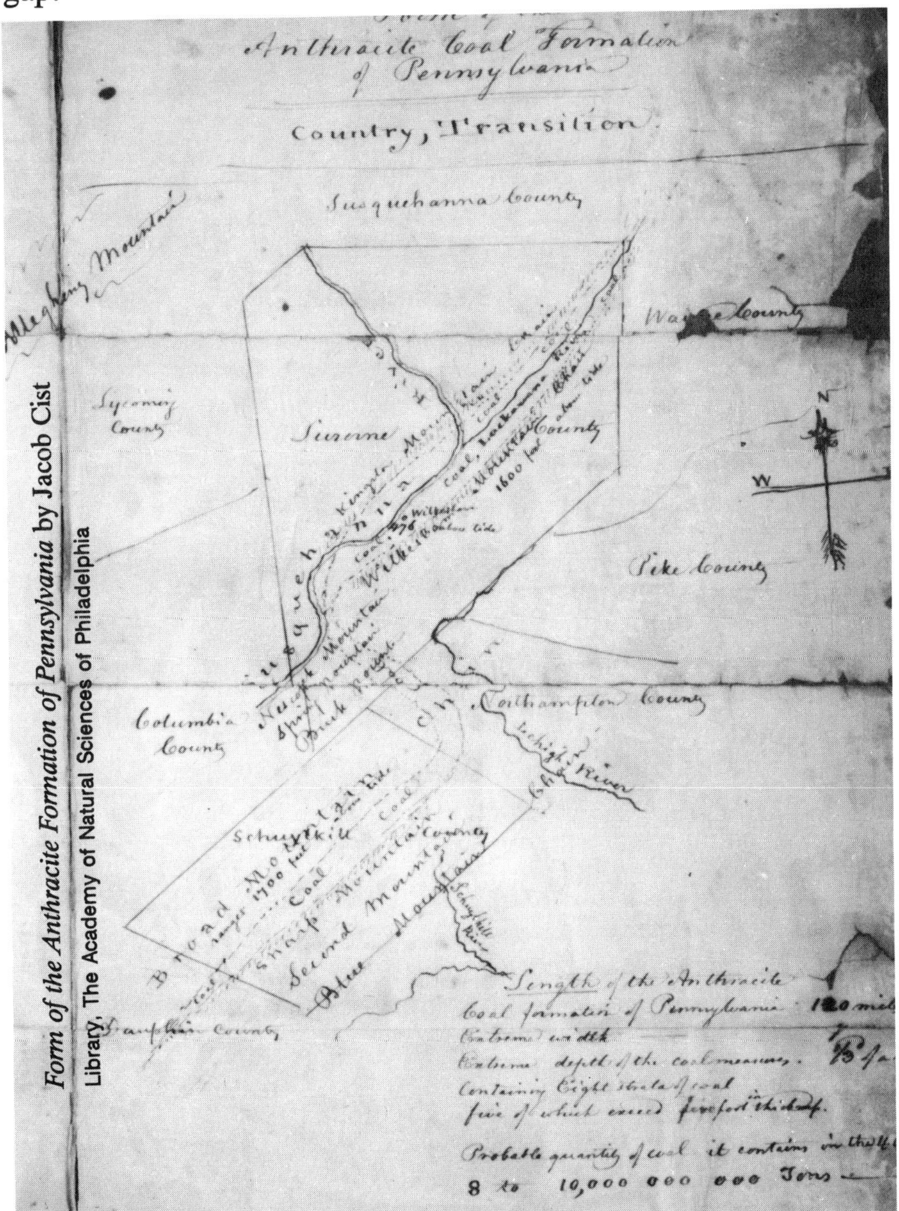

Form of the Anthracite Formation of Pennsylvania by Jacob Cist

Library, The Academy of Natural Sciences of Philadelphia

Cist sought testimonials from everyone who had substituted anthracite for bituminous. All the testimonials were enthusiastic, claiming a longer burning, more intense heat which made wood less appealing as a fuel. The price was competitive with bituminous, and even wood, when all the man hours needed for coking the raw wood were included.

The scientist and merchant got another boost in his efforts when one of Cist's oldest anthracite customers, and a shareholder from the days when his father, uncle and John Nicholson had started up The Lehigh Coal Mine Company, Philadelphia inventor and industrialist Oliver Evans shared results with Cist. Evans experimented with anthracite for his steam engine at his Mars Iron Works and later notified Cist that he had "found that hard [anthracite] coal excelled other fuel for keeping a steady fire and running the engine."

When Robert Fulton returned to the United States with a Boulton-Watt steam engine in 1806, Evans' steamboat took a back seat. But Evans had laid the foundation for American engineers to modify British inventions for this country's specific needs.

Very soon, the English steam engine and American anthracite would be working together, a combination nothing could stop.

Cist realized the significance of the combination. By 1823, he had projected the future of Luzerne and the future of Lackawanna County by sketching major mining operations for Carbondale, what would one day become Scranton, and the Wilkes-Barre area. And shortly before his death on December 30, 1825, Jacob Cist began working with George MacCulloch, of New Jersey, to construct a railroad from the Wyoming Valley to New York City, even though a canal was being built from Carbondale, Pennsylvania to the Hudson River at Kingston, New York.

Chapter 26

"...if we could but move our goods about America in easy commerce."

Josiah White wrote in 1812, "We pay higher prices to foreign manufacturers than we would to our own if we could but move our goods about America in easy commerce."

White was convinced that water power was not only the answer to increased manufacturing capabilities but also the solution for cheap, commercial transportation.

He devoted his life to perfecting both in Pennsylvania: "Pennsylvania has several rivers which could provide both navigation and waterpower. If such waterpower were developed here, it would claim the attention of many new manufacturers. Pennsylvania's watercourses, if also improved by locking, would create a diverse and extensive navigation for all."

Starting in 1812, White thought first of improving the Schuylkill River as it approached Philadelphia. He helped form the Schuylkill Navigation Company, anticipating a canal extending as far north as Reading, to bring coal, timber and grain from around Pottsville to the Philadelphia market. But by the time the company became active, White had been forced out.

White, however, could not ignore his great engineering ability, nor his vision of what could be.

From his experience operating a nail mill at the Falls of the Schuylkill (River), White had been granted two patents in 1812. And both White and mill partner Erskine Hazard, a fellow Quaker, had come to recognize the true worth of "the black stones that burn" when the War of 1812 with the British had closed off soft coal imports from England as well as shipments from Virginia.

The anthracite coal used by White and Hazard had come from Jacob Cist's Lehigh Coal Mine Company at Mauch Chunk (now Jim Thorpe).

In 1814, White and Hazard issued the following testimonial:

"To Jacob Cist, Wilkes-Barre:

"A Certificate from White and Hazard Proprietary, with their extensive Wire Mill and Rolling Mill at the Falls of Schuylkill, five miles above Philadelphia.

"We have used Lehigh Coal in the heating of bar iron for rolling. We find it to contrast with Virginia Coal as follows:

"With Lehigh Coal it takes 5 bushels per day of twelve hours to heat and roll one thousand pounds of iron for wire.

"With Virginia Coal it takes 20 bushels for the same, and as well twice the time and twice the cost in labor for the quantity of product.

"White and Hazard

"Whitestown, Pennsylvania

"November 1814"

Given White's inquiring, scientific mind and a testimonial like that - you can easily guess what would happen next -

By 1818, White, with Hazard, and a newly formed friendship with George Hauto, had leased the property of the Lehigh Coal Mine Company and received approval from the Pennsylvania Legislature to lawfully "enter upon the Lehigh River, to open, enlarge, or deepen it...and to make and set up any dam, locks, or devices whatsoever, which they shall think most fit and convenient to be a good navigation downward." The project had to be completed within 20 years.

An Act To Improve The Navigation Of The River Lehigh was afterwards described by a legislator: "We granted them the privilege of ruining themselves."

On April 1818, White and Erskine were at Mauch Chunk, surveying the Lehigh from Easton north to Stoddartsville;

by January 23, 1819, White was advertising in Philadelphia newspapers that he was taking orders for Lehigh Coal;

three months later - in April 1819, White and Hazard had sold their mill property to the City of Philadelphia, because it included the rights to the water power of the Schuylkill River;

and on October 19, 1819, White was granted another patent - this time for a "Canal Lock, Hydraulic Type."

Issuing the equivalent of what would today be called a "Press Release," White wrote, in part, "Lehigh coal is purer by twenty percent than any other now in the market and it can be brought to market in Philadelphia at an attractive price.

"In addition, by the development of the Upper Lehigh, and using a contemplated railroad connection between the Rivers Lehigh and Susquehanna, the Western counties of New York will be closer by expense of transport to Philadelphia than to Albany. The New York Canal [Erie Canal], when completed will connect Lake Erie and the West to the Lehigh River via the Susquehanna and a waterway using the Seneca River."

White had turned to the Lehigh River as the place where he could put his dreams into action!

White, Hazard, and Hauto (who later disappeared) first formed The Lehigh Navigation and Coal Mine Company in 1818. Meeting little success in selling shares, the entrepreneurs split the one company into two parts: Lehigh Navigation Company and Lehigh Coal Company. In 1822, the creation of the Lehigh Coal and Navigation Company reunited them into one company again.

England's success with canals had shown the Americans the way to transport heavy, bulk items. White was committed to canal transport. Initial reports confirmed his judgment. In 1820, the Lehigh Navigation Company had been able to bring anthracite coal to Philadelphia to sell at $0.30/bushel, compared with $0.43/bushel for British bituminous coal brought to New York City.

White happily predicted that when the Lehigh canal was completed, coal would be available in Philadelphia at $0.20-.25/bushel.

It looked as though White had found the answers for anthracite coal advocates - economically cheaper and easily delivered to the consumer.

But the man with vision had no intention of resting on his laurels. Jacob Cist and the northern coal field were not to be forgotten. White wrote in 1821:

"The Lehigh River presents itself as one of the most important links in the chain of a general inland navigation, approaching to within twelve miles of the Susquehanna at Wilkes-Barre..."

From the Lehigh, White could go anywhere.

But first, White, with William Wurts, brother of Maurice Wurts, set out on a trip to inspect the possibilities of a canal for the Wurts from their landholdings along the Lackawanna River.

They began their journey by calling on Thomas Morris, president of Schuylkill Coal Company, and learned that only fifty percent of the Schuylkill's stock had been sold. The two travelers then wandered all over northeastern Pennsylvania and into New York State as far as the Hudson River.

On July 9, 1823, White observed, "Found coal in a mass on the Lackawanna River. Coal appeared pervasive in the country adjoining the Lackawanna ."

And a week later, White "examined the Lackawaxen River," reporting on July 18, "...started in the rain for Milford and down the Delaware from near Shohola Falls. The range line of the hills favored a canal from Shohola Falls to Shinglekill on the Pennsylvania side, or a similar distance on the New York side." (For this portion of his trip, White was traveling in Pike County, erected from Wayne County in 1814.)

So it was White who helped lay out the route for what was eventually to become the Delaware & Hudson Canal.

White also continued to work on improving transportation within the mountains. Between April and May of 1827, White engineered the first railroad transportation in the country. To bring the coal from the mine near Mauch Chunk, a turnpike-type road was laid from the mine to the top of the mountain overlooking Mauch Chunk and then down to the town on the bank of the Lehigh River. For rails, two strips of wood were laid on the road three feet and seven inches apart, extending the entire length of the road.

Solomon Roberts, White's nephew, recalled, "The wooden rails were strapped with iron bars one and one-half inches wide by three-eighths inches thick.

"I helped push the first cars up the track for the first experimental run and rode down in one of the cars."

The following year, Roberts had been replaced, "We have also erected a steam engine for the purpose of drawing up the wagons loaded with coal from whence they will descend by the effect of gravity to Mauch Chunk."

The era of the railroad was about to begin.

But until the problem of ascending steep grades was solved, the stationary steam engine and gravity would be the answer to transportation over the mountains. For northeastern Pennsylvania, and coal enthusiasts, the gravity railroad was the solution.

Chapter 27

"...took his place alone, and set in motion the first locomotive ever run in the western hemisphere."

"...and soon I was out of sight in the three miles' ride alone in the woods of [northeastern] Pennsylvania."

It was called "Canal Mania" and there wasn't a self-respecting state in the union that hadn't had a "go" at it by the 1830s. Wisconsin, a western wilderness up to the time it became a state in 1848, had already commissioned a canal route with an accompanying congressional land grant by 1846.

Inland navigation, a euphemism for canals, expanded to include waterways for steam and sailing vessels, as technology developed in the 19th century. But it was primarily a call for a canal.

The two earliest canals of any consequence were the Middlesex and the Santee & Cooper Canals. The Middlesex connected Boston with the communities on the way to Lowell, Massachusetts, and on to Concord, New Hampshire. It opened in 1807. The Santee made Charleston, South Carolina an important port city. It opened in 1800.

The most famous, sometimes called the Grand Canal, was the Erie, connecting Lake Erie with the Atlantic Ocean. One of its earliest proponents was General George Washington, who viewed the area in 1783, although the Erie was not built and opened until 1817 - 1825. The attraction of the Erie was that it was the easiest way of connecting all the western part of the United States with Atlantic seaboard communities. It took advantage of the Mohawk River Valley, the only cut in the Applachian Mountains between Florida and the Saint Lawrence River in Canada. As an engineering feat, it was the length of the Erie that captured the imagination - 363 miles. (The Middlesex and Santee were each about 25 miles in length.) The rise and fall of 565 feet in elevation over the 363 miles averaged less than two feet per mile with just one major incline of 76 feet, at Lockport, New York.

And while the Erie was building, Josiah White was on the Lehigh River in Pennsylvania showing that a canal could succeed with a rise in elevation of 364 feet over the very short distance of 34.6 miles from Easton on the Delaware River to Mauch Chunk. White also showed that a canal from the anthracite coal region could do what it was expected to accomplish - it could bring out tons and tons of coal at a very low transportation cost.

This was the difference between the canals from northeastern Pennsylvania and those built elsewhere in the United States: canals that touched the coal fields existed primarily to carry coal. When they carried other commodities, such as timber or grain, it was like a windfall - extra money.

On canals like the Erie, as great as their tonnage traffic was, there was very little "steady trade." The Erie was dependent on farmers and manufacturers, and immigrants "going west." The Santee was perhaps most similar to the anthracite canals, as its purpose was to bring cotton and rice from the inland plantations.

Maurice and William Wurts had gotten their coal land along the Lackawanna River - near the western boundary of what had originally been Connecticut's The Delaware Company purchase (from the Delaware River to ten miles east of the Susquehanna River). First, the Wurts tried dragging a sledge loaded with anthracite over Moosic Mountain to the headwater of the East Branch of the Wallenpaupack River. From there, a succession of rafts, wagons and arks would take the coal to the Lackawaxen River, then the Delaware River, until finally the Philadelphia market was reached. But this was too complicated for mass tonnage delivery, so the Wurts looked about for another solution.

They found it when they studied the map Jacob Cist had sketched for them. Cist had marked coal further up the Lackawanna. The Wurts brothers next purchased land from Lord John Russell, of England, through his agent Joseph Priestley. Russell was the owner of several tracts in the area we know now as Carbondale, in Lackawanna County.

Russell was represented in the United States by Joseph Priestley, the British chemist who discovered oxygen. Priestley learned much about electricity from Benjamin Franklin; later, because of religious and political differences, he left England and settled in Northumberland, Pennsylvania, at the juncture of the East and West Branches of the Susquehanna River. Priestley's income was derived from his writings and from representing Russell.

And at William Wurts' request, Josiah White toured the area in 1823, showing how a canal route was feasible - but to New York City markets, not to Philadelphia.

If a canal was possible, then the die was cast. From 1823 to 1825, Maurice Wurts' Pennsylvania authority "To improve the navigation of the River Lackawaxen" metamorphosed from Lackawaxen Coal Mine and Navigation Company into a corporate body, The President, Managers and Company of the Delaware and Hudson Canal Company, popularly known as the D&H.

The brothers asked the engineer Benjamin Wright to come and build their canal. Wright was just completing his work as chief engineer of the Erie Canal. Wright accepted the job, and came to northeastern Pennsylvania, first sending several of his assistants from the Erie Canal, including John Jervis. The final plan for the D&H Canal called for a beginning on the outside of the mountain wall encasing the coal, at a place on the Lackawaxen River we know today as Honesdale, in Wayne County. From there, the canal would follow the Lackawaxen, cross the Delaware River, and then turn north, ending on the Hudson River at Kingston, New York, a distance of 108 miles.

The Wurts had a tremendous task. For one, they would have to bring the coal from a valley, the Lackawanna, up and over a mountain top, an elevation of almost 2,000 feet above sea level, before even thinking about floating the bulk weight to New York. Secondly, they needed approval from New York State as well as Pennsylvania; their company would have to be approved by two legislatures, not just one.

With the Mayor of New York City, Philip Hone, on the board of the D&H, and the Wurts' Quaker connections in Philadelphia, the canal project soon had sufficient clout in both states. Ground was broken for the canal portion of the project on July 13, 1825 at Summitville, in Sullivan County, New York.

To bring the coal from their mine to their canal, the Wurts decided on a series of stationary engines - the gravity railroad - to run from the coal mines at Carbondale to the canal terminus at Honesdale.

The D&H, then, would utilize the best of White's Lehigh Coal and Navigation Company - with improvements, if possible.

But the growing debate across the United States caused the Wurts to stop and consider - perhaps there was something to this railroad talk. So, in the fall of 1827, Horatio Allen, who had been an assistant engineer on the D&H project, left for England, with the blessings of John Jervis. The Wurts tasked Allen with investigating the railway, and if there was something to it, to bring home a railroad engine.

In England, Allen rode as a passenger on the Stockton and Darlington Railway; he met and talked with George Stephenson, the railway's creator, who, after Richard Trevithick, was the father of steam locomotion. The Pease Family, well known Quakers, was behind the Stockton & Darlington. It was the family's support for new technology that gave the Stephensons, father and son, the opportunity to set up the first steam locomotive (on a company-owned right-of-way) for public traffic on a regular schedule .

Allen then traveled to Stourbridge, England, where he ordered three steam locomotives of the Stockton & Darlington type from Messrs. Foster, Rastwick & Co.; one was the *(Stourbridge) Lion.*

After that, it was on to Newcastle, England, where Allen ordered the *America,* the same design as the *Rocket* that would win the Rainhill Trials in October 1829, from Messrs. Stephenson & Co. After the*America* arrived at New York City, it was displayed there with a full head of steam on May 27, 1829. The cost for all four locomotives was paid by the D&H.

By July 23, 1829, the *Lion* had arrived, too, and was shipped via canal to Honesdale, ready for its trial run.

"The first trip was made [on] August 8, 1829, in presence of a large crowd, which had assembled to witness the novel spectacle. Mr. Allen, declaring that if there was any danger he was ready to meet it, and that it was not necessary to sacrifice any other life than his own, took his place alone, and set in motion the first locomotive ever run in the western hemisphere."

Later Allen described, "When the steam was of right pressure, and all was ready, I took my position on the platform of the locomotive alone," and placed "my hand on the throttle-valve handle...The locomotive having no train behind it answered at once to the movement of the valve; soon the straight line was run over, the curve was reached and passed before there was time to think as to its being passed safely, and soon I was out of sight in the three miles' ride alone in the woods of [northeastern] Pennsylvania."

Everyone present that August day knew that they were watching a history-making event, that it was the first commercial railroad run in America.

But the Wurts brothers were sensible men, attuned to the realities of the world.

They recognized that "canal mania" was at it highest; that more than $100 million was invested by almost all the states and entrepreneurs in building canals; that the people of northeastern Pennsylvania had voted for the legislators at Harrisburg who would approve a state canal from Philadelphia to Pittsburgh in return for the North Branch Canal that would go all the way up the Susquehanna River - so that coal could leave the Wilkes-Barre area; and they knew they, themselves, had just spent over $1.3 million to build a canal from the Hudson River to Honesdale.

So quietly, very quietly - because the engine was too heavy for the track, it was said - the *Lion* was moved to the D&H maintenance shops - until forgotten - and then dismantled.

And the very level-headedness that the Wurts displayed in weighing in on the side of canals exacerbated canal believers. With the same kind of level-headedness, D&H management ran "a tight ship," thereby making the D&H about the only canal company with genuine profits.

Thus, all who wanted to build a canal needed only to cite the success of the D&H to find backers. The truth of the matter was, however, that canals almost always lost money.

But while canals seemed to have a natural affinity for balance sheets with red ink, they became the life-blood of America, tying the new nation into a unified country.

If Americans had not built canals, had instead patiently waited for railroad technology to develop to the point where it could perform the same tasks as the canals did initially, then the first half of the 19th Century would have passed into oblivion.

For the United States, the first half of the 19th Century was the good start - when Americans reached the gold of the California hills, when 20 more states were added to the union.

Chapter 28

But why would anyone want our goods?

"Now the coal business put on a different face. It [anthracite] could be used in private houses for fuel in place of wood, and would burn all night and keep a house warm in winter time."

Yes, indeed - who would want the commodities that northeastern Pennsylvania had to offer?

Land for development, for pioneer settlers to build towns and cities? There was land everywhere in the new United States of America.

Mature trees for timbering, stretching as far as the eye could see? Every state in the union had that.

Unusual black stones that could burn? Ah, only northeastern Pennsylvania had that.

(Pennsylvania has about 95% of the world's anthracite coal supply. Small deposits in New Mexico and Rhode Island, USA, and Wales, in Great Britain, are generally not commercially viable. Other anthracite areas currently mined include what was formerly the USSR, North Korea, China, South Korea, and Germany. But only Pennsylvania with its vast reserves could mount the kinds of commercial operations that existed in the 19th Century.)

"Coal" is a solid fuel of vegetable origin. (Oil is also of vegetable origin, but is a liquid fuel.) It is the word for a carboniferous substance formed over 250 million years ago. Coal is further subdivided into four major classes: *peat,* with low carbon content and high moisture content; *lignite and subbituminous*, a little more carbon and a little less moisture content; *bituminous (soft) coal,* still more carbon and even less moisture; and *anthracite (hard) coal,* almost all carbon with very little moisture.

World geology indicates that most of Planet Earth had a tropical climate at one time, covered primarily with lush vegetation growing in swamps. This time span has been labeled the Carboniferous Period of the Paleozoic Era, 345-250 million years ago.

Plant life, leaves and whole plants, as they died and fell to the ground, decomposed, throwing off oxygen and hydrogen (moisture). Like a compost pile of today, the remaining deposit was mostly carbon.

From 250 million years ago to the Age of Man (the last 10,000 years), mud and sand deposits, volcanic heat, and movements in the earth's crust (mountainous upheavals) pressed down on what was left of the dead vegetable matter, compacting the carbon and whatever moisture remained into what is now labeled as various classes of coal, depending on location.

The mountain upheaval in northeastern Pennsylvania, identified as the Appalachian Revolution, must have been extremely convoluted and exerted tremendous pressures, because it is only in northeastern Pennsylvania that a particular kind of anthracite coal, of a high carbon content with a low sulphur residue, is found.

In a comparison with bituminous (soft) coal, anthracite (hard coal) stacks up as follows: carbon content is generally identified as between 85-90% compared to 70-85% in bituminous. Anthracite's moisture content is listed as less than 15% whereas bituminous is listed as less than 20%. Anthracite can deliver approximately 22-28 million BTU (British Thermal Unit) per ton while bituminous ranges anywhere from 19-30 million BTU per ton. Anthracite ignites between 925°-970° Fahrenheit and bituminous ignites anywhere betwen 700°-900° Fahrenheit.

Because anthracite coal is almost all carbon, with only 4.3% volatile gasses, it is much, much slower to ignite - hence its first name: "the black stones that burn."

Bituminous coal had been in use in England as early as the 1600s. Its low ignition point had helped to define "coal." So when the black stones were found in northeastern Pennsylvania, everyone thought - coal. But then how could it be coal - when the black stones were so hard to ignite?

And there the matter stood: some who knew about the black stones were convinced there was something to them; others, who were more interested in the immediately practical - farming - would readily sell their land in northeastern Pennsylvania and move on, westward, to Ohio, to Indiana, eventually crossing the Mississippi River.

And then, there were a few hardy souls, who were absolutely convinced the black stones had a real value.

Brothers Obadiah and Daniel Gore were the first among the latter group. The Gores came from Connecticut in 1769. Before the end of the year, Obadiah had experimented with a few of the "Stone Coal" pieces in his blacksmith's fire, and was highly satisfied with the results.

Other blacksmiths built on Gore's discovery until there was general acceptance among iron workers that a fire made of "stone coal" was a better and hotter fire for working iron. "But such coal could be made to burn only by having a blast of wind from a bellows forced through it from below."

So, while a use had at last been found for "the black stones that burn," it was a limited use, restricted to forges and blacksmith shops working in a special way.

And then, Jesse Fell came to northeastern Pennsylvania.

When Jesse Fell came to Wilkes-Barre, he came with his family, his wife and four children, in 1785. They were Pennsylvania Quakers from Bucks County, north of Philadelphia.

The Fells opened a store and tavern in the heart of Wilkes-Barre. A likeable fellow, Jesse was soon accepted into valley life, and was elected sheriff of Luzerne County four years later in 1789.

His rigid Quaker background gave way to the exigencies of the situation, too, and it wasn't long before Fell "became prominent in military affairs." He was named brigade inspector in 1793.

"On the morning of the first parade of his brigade," Fell went out on the back porch of his house to do a little practicing. Dressed in full regimentals and with sword drawn, he shouted, "Attention, Battalion! Rear Rank, three paces to the rear, march!" and tumbled down into the cellar steps.

Who could resist such enthusiasm? The Connecticut settlers turned Pennsylvanians knew a good man when they saw one. Starting in 1798, Fell served as an associate judge of Luzerne County for 32 years.

Before his court appointment, the enterprising "backslider to the peace principles of the faith of his forefathers" had been active in many other matters, including a nailery, a forge limited to making nails. Fell experienced firsthand the potential power of the black stones. From then on, Fell never gave up trying to unlock their secret.

Then, at last, February 11, 1808, Fell succeeded.

He had been experimenting for so many years without a breakthrough that he eventually confined his experiments to the nighttime, when no one was around to see him try "one more time." On the evening of February 11, Fell tried a new grate design, elevated about eight inches above the hearth with iron rods spaced several inches apart. On the grate's bed, Fell first placed kindling wood, then the black stones on top. He lit the kindling wood, expecting, hoping, it would set fire to the black stones. He fell asleep, waiting. When he awoke, towards morning, he awoke to a warm, cheery room, brightened by an intense red glow from the fireplace. He had succeeded; he had gotten the black stones to burn!

Later Fell wrote, in a classic understatement: "Made the experiment of burning the common stone coal of the Valley in a grate in a common fire place in my house, and find it will answer the purpose of fuel, making a clearer and better fire, at less expense than burning wood in the common way. - Borough of Wilkes-Barre, February 11, 1808."

"Now the coal business put on a different face. It [anthracite] could be used in private houses for fuel in place of wood, and would burn all night and keep a house warm in winter time.

"Every family in Wyoming Valley, within a very short time, had a blacksmith make them a grate and a mason set it up in their fireplace every winter to burn this stone coal. A good fire could be kept all night with it, and comfort to a wonderful degree was increased in cold weather in their log houses."

And a year later in neighboring Hanover Township, young James Ross would ever after remember how "Crisman, the owner of the Red Tavern, opened a door and said to him, 'See, little boy, how nice the stone coal burns.'"

Now, the consumer had an intense need for "the black stones that burn" - for anthracite coal.

Chapter 29

Whatever the reason, large areas of land around Pottsville were still available as the 1820s' decade neared its end.

The first noticeable effect of Jesse Fell's discovery was a renewed vigor to bring anthracite to market. However, the increased activity was not seen until after 1815 and when it did begin, it was in the Mauch Chunk (now Jim Thorpe) and Pottsville areas, south of the Wyoming Valley. For it was at those places that Philadelphians could easily access the suddenly valuable and useful black stones with clear land titles and ready river transportation to bring the anthracite to their doors.

The Philadelphians' first thoughts turned to Pottsville because Jacob Cist, with his shares of Lehigh Coal Mine Company, was in possession of coal lands close to the Lehigh River at Mauch Chunk.

Josiah White had been one of the Philadelphians to consider Pottsville. By 1813, White, with Joseph Barnes, John White,

Timothy Abbott, John Conrad, and Conrad Carpenter, had formed The Schuylkill Navigation Company to build a canal from Reading to Philadelphia. (Reading is approximately 35 miles southeast of Pottsville on the Schuylkill River.)

But the struggle for control of the Schuylkill Company had so disenchanted White (who lost the presidency to Cadwalader Evans) that he had turned, in desperation, to utilize the Lehigh River instead; White would not be stopped. He first leased Lehigh Coal Company lands from Cist and then formed the Lehigh Coal and Navigation Company. On the Lehigh River, White was free to operate as he saw fit; the result was that his coal began arriving in Philadelphia four years before boats came from Pottsville by way of the Schuylkill Navigation Company.

The land around Pottsville, designated Schuylkill County in 1811, was rich in anthracite. But the exact locations of the outcroppings seemed to be a deep secret. While there was interest in coal land ownership, it may have been the Schuylkill Navigation's protracted efforts to get its transportation system going in the 1820s that acted as a damper on a major land rush.

Whatever the reason, large areas of land around Pottsville were still available as the 1820's decade neared its end.

All that changed beginning in 1829. A Philadelphia Quaker family, the Samuel Wetherills, acquired 2,000 acres west of Pottsville during that year. At the beginning of 1830, Stephen Girard, a Quaker originally from Josiah White's hometown (Mount Holly, New Jersey), paid $30,000 for 28,460 acres immediately north of Pottsville. Girard had moved from New Jersey to Philadelphia, as had White. But Girard had remained in Philadelphia, and grown wealthy from his mercantile and banking activities.

Shortly after the purchase, in the summer of 1830, Girard had his Schuylkill County property surveyed. Surveyors reported finding many signs of coal. Girard, however, prevented the mining of the land until 31 years after his death, when the land and its coal formed an important part of his estate, providing funds for his philanthropic trusts including Girard College.

While the Girard Trust lands lay idle, Henry Charles Carey and Isaac Lea, two brothers-in-law of one of Girard's surveyors (Thomas Baird), decided to invest heavily in Schuylkill County, and by 1835 had acquired over 42,000 acres.

Lea, a Quaker, had originally come from Wilmington, Delaware, to join the Philadelphia publishing house owned by Carey's father, Matthew Carey.

These three groups, Wetherills, Careys, and Girard, would account for almost all of the western portion of the southern anthracite coal field. The greatest significance of the owners was that they were all based in Philadelphia, and all directives emanated from the City of Brotherly Love. It was "absentee landlordism" with all its ramifications.

At first, however, the "absentee landlord" relationship worked to the advantage of all.

Both Wetherills and Careys included very large extended family memberships for operating and controlling the coal mining. Junior members of the families were dispatched to Pottsville and surrounds to live, if not permanently, at least for several years.

Lea himself was at heart a scientist. While coal lands represented money to him, they also satisfied his personal desires. In 1839, he published a report on finding a dinosaur footprint in the red shale of Sharp Mountain at Pottsville. His find of "the oldest case of an air-breathing animal on record" held up until 1885. Up to that time, geologists in both America and Europe referenced his work.

The proclivity or natural tendency of the "absentee" landowners was to reflect the activities of Philadelphia in their decisions when dealing with their Schuylkill County properties. This brought them to associate, and have as business partners, men who were interested in their interests. Burd Patterson, a man with ideas, was one.

Patterson had succesfully established the first wool-spinning factory in Philadelphia, helping to turn the Manayunk section of the city into a major textile center of the United States. From Philadelphia, Patterson moved to Pottsville and involved himself in coal activities. The coal towns of Ashland, Shamokin, and Mahanoy City had their beginnings in Patterson's dreams.

An entrepreneur, however, will at times work with someone else's money. So it would be only a matter of time before Patterson teamed up with the Carey and Lea family. Once together, Carey men and Patterson took anthracite mining below the water table for the first time. (The water table is the level below which earth's voids are filled with water. To mine below the water table meant some arrangement would have to be made to drain mine tunnels of water.) Soon specially built pumps and steam engines were in place, and tunnels, or shafts, thousands of feet long were carrying miners deep into "the bowels of the earth." This, in turn, led to the development of a sophisticated system for mine ventilation.

In the 1830s, scientific advances were very much on everyone's mind in Philadelphia, probably because the Franklin Institute was barely five years old. (The Franklin Institute is in existence today. As when it was founded, it continues to serve as a museum of science and industry - with many hands-on displays, a planetarium, research center; a publisher of scientific articles, and a promoter of scientific research.)

In 1825, the Franklin Institute offered a gold medal to the first person who could make a commercially viable quantity of iron from iron ore using only anthracite coal. And in 1836, an article in the *Journal of the Franklin Institute* described a new method for breaking anthracite into usable pieces - the beginning of the "breaker."

(Prior to the development of the breaker, anthracite coal, as it came out of the mine, was placed on a large steel plate and broken into smaller pieces by a sledgehammer. The breaker is a structure up to a 100 feet high. Railroad cars, filled with anthracite from the mine, enter the breaker at the top, dropping their cargo onto huge rollers. The continuously revolving rollers break up the coal, which falls onto progressively smaller size screens, automatically sorting the coal into "buckwheat," "chestnut," "pea," and other sizes for commercial consumption.)

Spurred on by activities at the Franklin Institute, because they were publishers with scientific minds, and because they had Patterson at Pottsville, Carey and Lea looked at Pottsville from a Philadelphia viewpoint, and concentrated on putting Pottsville at the forefront of mining activities.

Chapter 30

"On that date it was conclusively settled that iron could be smelted with Pennsylvania anthracite coal, exclusively by the use of hot blast."

The race was on! Who would succeed? Who would be able to smelt iron ore on a commercial basis, using only anthracite coal?

A New York City Lutheran clergyman, Dr. Frederick Geissenhainer, obtained a patent on December 19, 1833, based on work beginning in 1805 using both hot and cold blasts at his Valley Furnace at Silver Creek, northeast of Pottsville near New Philadelphia. While successful, Geissenhainer could not keep the blast going due to other furnace deficiencies.

In the end, success came to Welshman George Crane, owner of Yniscedwyn Iron Works, Swansea, South Wales, Great Britain. The *Journal of the Franklin Institute* published Crane's patent in 1838. Crane's British patent had been recorded on March 28, 1837, but when Crane applied for a U.S. patent in addition to his British one, Geissenhainer's patent, already recorded, stopped him. The enterprising Welshman then purchased Geissenhainer's patent rights from his heirs, made some improvements, and received a U.S. patent on the improvements in November 1838.

Erskine Hazard, Josiah White's original partner and also a large shareholder in Lehigh Coal and Navigation Co. of Mauch Chunk, arrived in Wales shortly after Crane received his patents. The Lehigh Company wanted Crane's rights.

Crane, instead, gave the Mauch Chunk firm the services of David Thomas, Crane's superintendent at the Yniscedwyn Works, and royalty rights.

The Lehigh Crane Iron Company was formed, and Thomas, with his family, came to Biery's Bridge on the Lehigh River (now Catasauqua) to set up and operate a hot blast iron furnace for the Lehigh Coal and Navigation Company.

But...

Thirty-five miles to the southwest, over in Pottsville, under the aegis of the Carey/Lea family, Burd Patterson had begun building an iron furnace, Pioneer Furnace, locating it on an island in the middle of the Susquehanna River. (Pottsville founder John Pott had erected a furnace there originally in 1807.)

Patterson brought in William Lyman of Boston to do the actual building, and Benjamin Perry of Pottsville as ironmaster. Like the Lehigh group, Patterson and Lyman were operating under Crane's patents.

Definitely, the race was on!

Lehigh Crane ran into complications first. Because it was trying to be the best, David Thomas' specs (specifications) called for a part - five foot diameter "blowing cylinders" - that was too large for shipment from England.

While Thomas sat on the banks of the Lehigh River waiting for the cylinders to be specially made in Philadelphia, he "received a communication from this gentleman [Lyman, of Pottsville] by the president of the Lehigh Crane Company [Robert Earp; White and Hazard were members of the board of directors], for whom I was building the first furnace at this place [Catasauqua]. This letter urged me to come to Pottsville.

"I visited him [Lyman] in August, 1839, and furnished him with plans of inwall, bosh, hearth, etc., and continued to visit him about once a month until the furnace was completed."

By then, the long-awaited cylinders had been shipped up the Lehigh and Thomas concentrated on his work at Lehigh Crane Iron.

But it was all the help Lyman needed. The Pottsville furnace was using four foot diameter cylinders; there were no delays, and in the end the blast was blown in October 19, 1839.

Three months later, on January 18, 1840, "an experiment was consummated in Pottsville which had a wide influence upon both state and nation.

"On that date it was conclusively settled that iron could be smelted with Pennsylvania anthracite coal, exclusively by the use of hot blast."

At a victory celebration held that evening in the Mount Carbon House, Nicholas Biddle, up from Philadelphia, toasted the jubilant winners:

"[To] Old Pennsylvania - her sons like her soil: rough outside, but solid stuff within; plenty of coal to warm her friends, and plenty of iron to cool her enemies."

Biddle, also of Quaker descent, would soon be a major player in the southern coal field. President since 1822 of the national bank, Second Bank of the United States, Biddle's fight with U.S. President Andrew Jackson in the 1830s had prevented renewal of the bank's charter. To overcome this obstacle, Biddle secured a state charter in 1836, and the bank became Bank of the United States of Pennsylvania. At the time of his toast, Biddle had just resigned from the bank to join the Girard Trust in establishing Girard College, whose income would depend on developing the Stephen Girard coal fields north of Pottsville.

(Biddle's home, called "Andalusia," on the banks of the Delaware River north of Philadelphia, was given to the National Trust for

Historic Preservation in the 1970s and is now a museum open to the public.)

Amazing, what an unusual bit of history was played out in Mauch Chunk and Pottsville in 1839 and 1840. Such altruism! What captains of industry today would help their arch competitor?

The key lay in Biddle's toast:

Pennsylvania's sons were the Solid Stuff (the Quaker influence);

Plenty of coal to warm our friends (a monopoly in home heating because of Jesse Fell's discovery);

Plenty of iron to cool our enemies (a monopoly on making better iron).

Please, don't be too cynical. The altruism was genuine. In the middle of the three month trial period, as it was daily becoming more apparent that the Lyman/Carey/Patterson group would win the prize, the Pottsville newspaper *Miner's Journal* wrote:

"*We shall now soon find that all our railroad iron, besides the immense quantity used for other purposes, will be made at home, without a servile dependence on other countries.*"

Pennsylvania assets and her "solid stuff" men were guarantee that the American Revolution for freedom could not fail.

(Of course, when you have a monopoly....

But the evils of a monopoly, just like the damage from absentee ownership, would not come until afterwards.

So it really was altruism, but in the form of enlightened self-interest, as all the chips were stacked equally between both parties at the beginning of a rapid and dramatically expanding marketplace.)

Once the technology existed, anthracite iron furnaces popped up everywhere. A major find of iron ore in Columbia County caused the erection of furnaces and mills around the junction of the East and West Branches of the Susquehanna River. Danville became a center. The North Branch Canal from Wilkes-Barre brought the anthracite to operate the furnaces. (Bituminous coke furnaces, such as in the western part of Pennsylvania at Homestead, did not begin to have an impact until after 1890.) But as active as Columbia County was, Schuylkill and Lehigh Counties dominated the market until 1900. For example, in 1873 anthracite furnaces were responsible for 46% of all iron made in America. Of that 46%, almost three-quarters of it was produced in Pennsylvania.

The greatest power of the monopoly, as Biddle's toast implied, was immediately before the start of the War Between the States (Civil War - 1861-1865). There were 120 anthracite iron furnaces in the United States in 1858; of that number, 92 were located in Pennsylvania. The others were divided: New York-15; New Jersey-5; Maryland-4; Massachusetts-3; and Connecticut-1.

When Confederate General Robert E. Lee and his troops entered Pennsylvania in 1863, they were definitely on the offense. They also knew that at Gettysburg they were only 100 miles from the North's iron and coal center. The Battle of Gettysburg on July 1-3, 1863 is considered to be the turning point of the Civil War. When the Confederate troops turned back, it was at a cost of roughly 3,000 lives, 13,000 wounded, and 5,000 prisoners on each side. The South never again invaded the North - and coal and iron continued to pour out of the northeastern Pennsylvania mountains in an unstoppable flow.

Chapter 31

The cry for improved navigation on the Susquehanna grew ever louder after Jesse Fell demonstrated the very practical application of anthracite as a home heating fuel. But the Susquehanna River was not like the Lehigh or Lackawaxen Rivers; it was a main artery of travel.

Improved river transportation was the focus of "best available technology" for those having to ship bulk items. The communities in northeastern Pennsylvania along the banks of the great Susquehanna River concentrated on improving river conditions to send their products to market.

Water transport on the Susquehanna could easily carry goods to Baltimore and Washington, D.C. markets, and from the Chesapeake Bay - everywhere else.

The first coal leaving northeastern Pennsylvania, shipped to Carlisle at the time of the American Revolution, was from Wilkes-Barre. Later, Wilkes-Barre, as a source of anthracite coal, expanded into an area including Kingston, Plymouth, and Pittston.

From these and intermediate points, coal was floated down the Susquehanna in Durham boats or Susquehanna coal arks, carrying anywhere from five to 60 tons of coal.

(Durham boats were originally built by Durham Iron Furnace, in Bucks County near the Delaware River. Starting in the 1730s, the company used the boats to carry its products to market. The boat design proved so popular that it was soon adopted for use along the whole length of the Delaware for all kinds of bulk shipments. After that, Durham boats appeared on the Susquehanna, Schuylkill, Mohawk, and St. Lawrence Rivers. The boat was approximately 8'x60' with a capacity of five tons.

The Susquehanna coal ark was used by Abijah and John Smith for shipping coal from their mine at Plymouth, starting around 1807. It was probably derived from an ark built by Mr. Kryder in 1792 for use on the Juniata River. The boat measured 16'x90' with a 4' depth. Its capacity was 60 tons. It proved to be a popular design, too.)

The problem with river shipping was that the boats and arks had difficulty in navigating rapids and rocky areas. Of every shipment that left the Wyoming Valley, about one-third failed to reach tidewater.

The cry for improved navigation on the Susquehanna grew ever louder after Jesse Fell demonstrated the very practical application of anthracite as a home heating fuel. But the Susquehanna River was not like the Lehigh or Lackawaxen Rivers; it was a main artery of travel.

Pennsylvania would not grant an exclusive charter for improving the Susquehanna as it had to Josiah White and the Wurts brothers for the Lehigh and Lackawaxen Rivers.

Instead, the Commonwealth would assume ownership - and responsibility for constructing a canal along the banks of the Susquehanna River - as part of its internal improvement program.

Tax dollars would build the canal.

(Tax dollars would also mean years of delay.)

Pennsylvania had an eye on what was happening with New York State and its sponsorship of the Erie Canal. If New York accessed the Great Lakes and had a monopoly on western travel, then Philadelphia's port could go into a decline as well as all of Pennsylvania.

The Potomac Company of Virginia (originally started by George Washington in 1789) was thinking the same thoughts, too, as the Erie Canal neared completion; and in 1828, the U.S. Congress subscribed to 10,000 shares in the newly-formed Chesapeake and Ohio Canal Company, successor to The Potomac Company, to build a canal from Washington, D.C. to Cumberland, Maryland.

Shortly before Jacob Cist died in 1825, he saw political advantages for northeastern Pennsylvania if support was given to Philadelphia, which was interested in building a canal across Pennsylvania to Ohio.

Although by this time Cist personally had come to believe that the future was with railroads, he acted to throw the weight of northeastern Pennsylvania votes behind Philadelphia, so that once the east-west Main Line Canal was begun, a Northeast Extension Canal (to Wilkes-Barre) couldn't be far behind.

(A note of interest: the City of Baltimore, like Cist, believed the future was with the railroad and did not give way to "canal mania," thus laying a foundation for the first commercial railroad system in the United States - the Baltimore and Ohio Rail Road.)

Wilkes-Barre desperately needed some form of mass transportation. Philadelphians had invested heavily to bring anthracite and iron to market, as well as lumber and grain. But the investments had always been made outside the center of activity of the old Connecticut claim. And what had at first been a natural advantage - the Susquehanna River - was fast turning into a major drawback because it would be a major undertaking to improve navigation on the river.

Cist's political trade-off worked. Matthew Carey, founder of the Philadelphia publishing house and father/father-in-law to Henry Carey, Isaac Lea and surveyor John Baird, was also president of the Society for the Promotion of Internal Improvements in the Commonwealth, known as the Pennsylvania Internal Improvement Society, which was organized in 1824. (John Nicholson and Robert Morris had formed the Society for Promoting the Improvement of Roads and Inland Navigation in 1789.) Carey's Society pushed for both an east-west canal, the "Main Line Canal," and a branch canal, the "North Branch Canal," to the northern coal fields.

But this was in 1825, shortly before Cist died, and several years before the Carey Family owned its own coal fields and iron furnaces at Pottsville. Gradually, the push to complete the North Branch lagged as activity around Pottsville increased.

Men from Wilkes-Barre stepped in to keep the idea of a North Branch alive, including the sons of Matthias Hollenback and Zebulon Butler. Perseverance succeeded; in December 1830, the first shipment from Shickshinny, south of Wilkes-Barre, reached Philadelphia by way of the North Branch, Union and Schuylkill Canals.

But it would take until October 6, 1856, another 26 years, to open the North Branch all the way. Then, goods, including anthracite coal, could be shipped from the Wyoming Valley either north to Athens on the New York border to connect with New York's Junction Canal and on to the Erie Canal, or south to Philadelphia, or Baltimore, or west to Pittsburgh.

Along the way, engineer William B. Foster, Jr., would come to work on the North Branch extension starting in 1839, bringing with him his younger brother, Stephen Foster. (*Today*, four Pennsylvania roadside historical markers trace Stephen Foster's path from Wyalusing, to Towanda and then Athens, in Bradford County. Foster's first piece, *The Tioga Waltz,* and later *Camptown Races,* were written at Towanda and Athens.)

By 1858, however, Pennsylvania wanted to be out of the canal business; it was losing too much money. A private company, the North Branch Canal Company, took over. Some of the directors of the new company were on the board of directors of the Junction

Canal Company, of New York. Some of the directors of the Junction Canal Company were on the board of directors of the North Branch Canal Company. Charles F. Wells, Jr. of Athens, George M. Hollenback and J. W. Hollenback of Wilkes-Barre, and John Arnot of Elmira, New York, were the interlocking investors.

But a combination of repairs, low receipts and finally, a flood on March 18, 1865, convinced the new owners that they should try something else.

Two days later on March 20, 1865, by emergency Act of legislature and governor, the company had a new name - Pennsylvania and New York Canal and Railroad Company.

Chapter 32

In all the hubbub of building canals and gravity railroads, there developed a forgotten land along the length of the Lackawanna River valley.

By the beginning of the 1840s, all of the major anthracite coal deposits had been identified; all the main transportation arteries leading out from Towamensing, the wild place - northeastern Pennsylvania - were in use, carrying anthracite to the rest of the country.

*Except for one area......*where coal was known to exist.

In all the hubbub of building canals and gravity railroads, there developed a forgotten land along the length of the Lackawanna River valley.

Unconsciously, Josiah White had commented on the state of affairs in the mid-valley when he toured the area with William Wurts in 1823, "At Isaac Vaughan's saw contracts for carting coal at one dollar seventy five the ton - via the Wallenpaupack and Paupack route. Found a gap through the mountain on the east face near Roaring Brook. Coal appeared plentiful along Roaring Brook." (Did Wurts tell White this had been the general path he and his brother had first taken in 1816 which had proved so disastrous?)

So, what would happen to the coal that lay buried along the Lackawanna? Would it be a forgotten treasure?

The Henry family had been producing ironmongers since 1776. Almost every generation had a William Henry, too. And the Henrys had always been members of the Moravian Church, settling first in the Lancaster County area. (Later Henrys would move closer to church headquarters, near Nazareth and Bethlehem.)

The William Henry of 1755 served under George Washington in Braddock's defeat at Ft. Pitt (Pittsburgh). This William was the kind of man who would save the life of an Indian, Killbuck, during the fighting. Later, in 1774 at the start of the American Revolution, Chief Killbuck met up with the next generation of Henrys. "Indian never forgets," the Chief reminded William #2. And after, besides kindnesses from Killbuck's tribe to Moravian missionaries, every member of Killbuck's family had "Henry" for a middle name.

It wasn't surprising then, that the third generation of William Henrys became a part of Henry W. Drinker's third attempt at constructing a railroad from the Delaware River, at the Delaware Water Gap, to the Susquehanna River at Pittston. William Henry's attraction for Drinker was his expertise in iron, vital to building a railroad. Drinker's 1826 railroad attempt included Henry as one of the charter members.

(Henry W. Drinker, with his brother Richard Drinker, had inherited 30,000 acres, known as Drinker's Beech, immediately southeast of today's Scranton known as Drinker's Beech in the 1830s from their father Henry Drinker, Jr., cashier of the Bank of North America at Philadelphia. Henry W. was a second cousin to brothers Henry and William Drinker, grandsons of Henry Drinker, the Elder, who owned 500,000 acres in Susquehanna and other counties and had been Samuel Preston's partner. All the Drinkers were Philadelphia Quakers.)

The William Henry of the 1820s was interested in branching out on his own, too, away from the family gun business, a large manufactory at Jacobsburg, near Nazareth.

Growing out of his work with Drinker, and the plans for a railroad across the Pocono Mountains, Henry and a nephew, John Jordan, Jr., obtained 800 acres and a mill site at Analomink, in today's Monroe County, about six miles north of Stroudsburg. They quickly established a forge and all the community buildings necessary to operate, including eight dwellings for their employees.

At the time, Henry's brother Matthew Henry, who had Catherine Furnace in Bushkill Township, Monroe County, was going to supply his relatives with all the pig iron necessary to operate the new forge. Soon, however, Matthew Henry was falling behind.

To provide iron for the forge, Henry and one of his brothers-in-law, John R. Wollee, with nephew Jordan, Jr., took a ten year lease, starting in 1832, on the Oxford Furnace, located across the river in Belvidere, New Jersey, near the Delaware Water Gap.

The Oxford Furnace had been in and out of blast since 1743. Cannonballs for the Continental Army had been made at the furnace during the War of 1776. But eventually, it had been shut down, lying idle for over 20 years, before Henry took an interest in starting it up again.

Henry had already established a reputation for himself as an ironmaster when he leased Oxford. In starting the furnace up again, he and his relatives made substantial improvements to the physical plant, rebuilding and enlarging it. Besides the renovations, Henry hired Selden Scranton in 1834 to be his assistant. Henry also experimented with improving the ironmaking process.

On May 24, 1835, Henry successfully used the hot blast technique for making iron in this country.

To make iron in quantity commercially, a furnace, about 35' high (the stack), is constructed. (David Thomas had insisted on a stack height of 45' for the first furnace at Lehigh Crane Iron Company in 1839.)

All of the ingredients necessary for obtaining iron from the ore are dumped into the furnace from the top of the stack. Furnaces would invariably be built into the side of a hill for easy access to both the top of the stack and the bottom, the furnace hearth. The ingredients used would be a source of energy (wood, charcoal, anthracite and/or bituminous coal, coke), the iron ore, and oyster shells or limestone (known as "flux."). Collectively, the ingredients form the "charge." The charge goes into the stack in successive, repeated, layers. A fire is lit, and air, in early furnaces by massive bellows, is forcibly blown through a "tuyere" (nozzle) from beneath to keep the fire burning. Once lit, or "blown in," it takes about 12 hours before the iron ore begins melting. A furnace can burn continuously from a few days to more than a year simply by adding more and more of the ingredients.

Iron furnaces of the 1800s were usually constructed of stone and lined with two layers of fire brick and a clay insulation. Furnaces were about 25' on a side at their base. A furnace stack is widest near its bottom, an area referred to as the "bosh." Beneath the bosh, in the "crucible" (a bowl-like depression), pure molten iron collects, waiting to be "drawn off" into a mold. All during the process, a hole below the bosh allows "slag" to be raked out and discarded. The purpose of the flux is to mix with carbon and other impurities in the iron ore to make a vitreous liquid, slag, that can easily be removed, leaving pure iron to run into the crucible.

The idea of pre-heating the blast of compressed air forced up through the bosh was first used in Scotland in 1828. James Beaumont Neilson patented *Hot Blasts for Furnaces* on February 18, 1829, based on his experiments at the Clyde Iron Works, Glasgow, Scotland. Neilson, however, did not use anthracite as part of the charge. (That was for George Crane to do in 1836-1837.)

While William Henry was not the first to use the hot blast, his was the first documented use in this country.

Chapter 33

For one, they were buying 503 acres at the confluence of Roaring Brook and Lackawanna River (the center of today's Scranton). Iron ore and anthracite coal had been found on the property.

In a series of events starting in 1836, William Henry embarked on a project that would forever change the "forgotten land along the Lackawanna River."

It all began when George Hollenback of Wilkes-Barre, son of the Wyoming Massacre survivor Matthias Hollenback, wrote to Joseph J. Albright, of Mauch Chunk.

In later years, Albright would recall that in 1836, "At the suggestion of George M. Hollenback I made the trip to Slocum Hollow for the purpose of examining the iron ore, coal, &c., with a view of purchasing from Alva Heermans the property (now Scranton) for $10 per acre. I took a box of the iron ore on top of a stage to Northampton County, where I was engaged in the manufacture of iron...I believe the box of ore thus transported was the means of attracting the attention of Messrs. Henry, Scranton, &c., to this tract. These facts are known and recognized by S. T. Scranton [Selden T.]..."

(Joseph J. Albright, with Jacob Rice and Samuel Templeton, had taken over the ownership and operation of Clarissa Furnace in 1834, renaming it Ashland Furnace. The furnace was located near Little Gap, at that time a part of Northampton County. Nine years later, in 1843, Little Gap became a part of the southeast corner of Carbon County.)

As Albright noted, his box of iron ore attracted Henry's attention, an easy thing to do as Albright and Henry were brothers-in-law. As noted above, Henry had only one source of iron at this time - from around his Oxford Furnace in New Jersey. Albright's box held out the bright prospect of an even larger iron reserve.

With renewed enthusiasm, Henry W. Drinker and William Henry convened a meeting in the city of Easton of the board of directors of their chartered, but as yet unbuilt, railroad company, the Susquehanna and Delaware Canal and Railroad Company. The Board also included: as its president, Lord Charles Augustus Murray, son of England's Earl of Dunmore (after whom Dunmore Borough took its name); and Edward Armstrong, from New York State. A common interest in hunting had originally brought Drinker, Murray, and Armstrong together.

William Henry's proposal was to reactivate an abandoned iron furnace on the property where Albright had obtained his box of sample ore. Henry maintained that once the furnace was operating, a thriving community would grow around it, and a railroad, their railroad, would flourish - an economic success.

The meeting at Easton ended, gaining Henry only one convert: not Drinker, not Murray, but Edward Armstrong.

Henry continued to operate Oxford Furnace, making improvements. It was not until 1839 that he returned to the Lackawanna Valley. But the property had been sold!

Henry "examined every rod of ground along the river from Pittston to Cobb's Gap to ascertain the most judicious location for the works." As far as Henry was concerned, and for what he had in mind, nothing could be found to equal the original tract of land. Finally in March 1840, by paying $6 more per acre, Henry and Armstrong were able to buy the tract of land offered to Albright in 1836.

What were they buying?

For one, they were buying 503 acres at the confluence of Roaring Brook and Lackawanna River (the center of today's Scranton). Iron ore and anthracite coal had been found on the property.

For another, they were buying the remnants of a forge and mill that had been lying idle for almost 15 years.

They were also buying an important piece of the area's history.

In 1798, brothers Benjamin and Ebenezer Slocum had purchased the land. It had been their sister, Frances, who became the "Lost Sister of Wyoming," when she was snatched by the Indians and carried away in 1778, not to be found until 1835.

The primitive grist mill drawing power from Roaring Brook was improved by the Slocums and a distillery was added, "frequent calls for whiskey coming from Easton, Paupack, Bethany, Montrose, and Berwick."

The following year, in 1799, the two Slocums started up a bloomery forge. (A "bloomery" is a primitive shop where small quantities of iron ore are melted in a cauldron to produce a pasty red-hot lump: the bloom. In iron making, bloomery iron is regarded as wrought iron, as distinguished from cast or rolled iron.)

The area soon became known as Slocum Hollow. But by 1826, Ebenezer had bought Benjamin's interest and the last iron was made. Two years later, Ebenezer retired. Gradually, Slocum Hollow fell into disrepair, waiting...

As with the forge at Analomink, Henry envisioned a whole community rising on the banks of the Lackawanna River and Roaring Brook. Edward Armstrong could see the same vision. Armstrong gave Henry the initial deposit money, $500, to secure the property at Slocum Hollow.

Then - Armstrong unexpectedly died. The administrators of Armstrong's estate withdrew support for Henry's project at Slocum Hollow.

Was Henry's dream to end?

Thanks to the Scranton brothers-NO!

Shortly after Selden Scranton came to work at the Oxford Furnace, his love for Ellen, William Henry's daughter, deepened, and the two were married.

Soon, Scranton had an ownership interest in Oxford Furnace and had brought in a brother, George W. Scranton, to join him.

These two Scranton brothers were intrigued by Henry's reports describing Slocum Hollow. A merchant in Belvidere, Sanford Grant, heard about the events surrounding Armstrong and the possibilities at Slocum Hollow.

On August 19, 1840, after a day and a half of travelling, the four men, with Henry leading the way, arrived at the Hollow. "Securing their horses under the shade of a tree, the party, amazed at the simple wildness of a country...moved down the bank of Roaring Brook to a body of coal whose black edge showed the fury of the stream...None of the party except Mr. Henry had ever seen a coal-bed before..."

With a pick, "large pieces of coal and iron ore balls lying in the vicinity were dug up."

To stand in Armstrong's place came Selden T. Scranton, George W. Scranton, and Sanford Grant. The real estate purchase was completed.

Afterwards, Henry "submitted a plan and an estimate for a blast furnace, with sufficient houses to accommodate the necessary workmen."

Philip H. Mattes, whom the Scrantons had come to know as depositors in Mattes' bank at Easton, was asked to join in the new venture.

Scrantons, Grant & Co. was formed; and on September 11, 1840 "the first day's work for this Company was done here [Scranton, Pennsylvania] upon the furnace."

Chapter 34

No one else in America had ever made that much "T" rail nor at that price. "To the Scranton Company, it was everything!"

The possibilities of Slocum Hollow became a mesmerizing phrase on which Selden T. Scranton, George W. Scranton, Sanford Grant, Philip H. Mattes, and William Henry hung all their hopes and dreams. It sustained them through their darkest hours.

Would it, in the end, become a false illusion, hypnotically leading them to an ignominious death?

So it would seem. For every step in the first days of the new venture met with failure.

The iron ore balls at the site of the furnace soon proved too expensive to mine.

Three miles away, on Moosic Mountain, a large ore deposit was discovered. This necessitated: (1) acquiring not only the land that had the iron ore, but also land between it and the furnace, almost 4,000 acres in all; and (2) which in turn required some means of bringing the ore to the furnace. The proprietors of Scrantons, Grant & Co. opted for a short railroad.

Soon, start-up capital was exhausted, thereby leaving "from the very outset an embarassing debt."

Could the visionaries succeed?

William Henry had moved with his family to the area as soon as the land had been purchased. Slocum Hollow became Harrison. Now, in the fall of 1841, George W. Scranton came to live beside Henry. This Scranton was the first to show that here was a man "whom no difficulties could discourage."

When the furnace was ready, a blast was started. But "owing to various causes incident to a new, wet, defective stack," it failed.

Then, a second effort failed!

Now, the furnace was altered, including major changes in the hot air ovens, machinery, and the addition of the services of a consultant.

"On the 18th of January, 1842, the furnace was blown in, amid mutual applause and congratulation."

A year later, Scrantons, Grant & Co. was making a superior iron but in very limited quantities. But regardless of the quantity, it was also a major effort to bring the product to market. There was almost no demand locally. To reach the cities, a 15 mile trip overland in wagons was necessary just to reach the first stage of commercial transportation - either to Carbondale and the D&H Canal, or to Port Barnum (Pittston) and the North Branch Canal on the Susquehanna River.

"The darkest period in the history of the partnership [Scrantons, Grant & Co.] was seen in 1842-3."

In 1843, the great visionary William Henry departed for Belvidere, New Jersey, and Oxford Furnace.

Selden and George Scranton remembered their first cousins, brothers Joseph H. and Erastus C. Scranton, in business in Augusta, Georgia.

Would they be interested?

Yes, they would - money was loaned.

Then George Scranton got the idea that if they concentrated on making nails, they could ship kegs more easily than iron bars; also, nails could put them directly in touch with the consumer market.

Selden Scranton was dispatched to New York City where he raised another $20,000 from John H. Howland.

Nail production began in 1844, but soon appeared to be heading for disaster. The nails were too brittle, one out of three breaking in use.

So, Selden Scranton came to live at Harrison and George Scranton returned to Belvidere and Oxford Furnace.

It was still a "touch-and-go" operation.

More changes were made. Joseph H. Scranton came up from Georgia to live in Harrison. Sanford Grant sold his interest to Joseph Scranton.

Joseph then talked with Joseph C. Platt of Connecticut and prevailed upon him to come to Harrison and take over Grant's work.

The year was now 1846 and the economic future of William Henry's vision was fast fading.

But out there, beyond the mountain walls surrounding Harrison nee Slocum Hollow, there was an economic future pulsating with life and vigor.

How could this other world affect the little band of iron makers on the Lackawanna River, who refused to give up? Who had a mountain of debts; who changed their company name as new investors came in; who rotated personnel, all in an effort to succeed?

The economic future that was pulsing with life was the railroad. It was almost the end of the 1840s, 15 years after the Baltimore and Ohio Railroad had started a regular schedule of operations with its Tom Thumb engine, and railroad tracks were being laid everywhere.

"Railroads...had begun to awaken unusual attention through the country [United States]. This naturally introduced among iron-men the subject of making railroad iron. The English ironmasters...predicted that this was a branch of trade the [Americans] would never blunder into...The monopoly of making the heavy 'T' rail could never pass from their [British] hands."

Some "T" rail had been made in the United States in 1845, but the bulk of the rails was imported from England.

News that the New York and Erie Railroad had laid track as far as Port Jervis, New York, on the east bank of the Delaware River across from Milford, on its way to Binghamton and eventually Buffalo, New York, reached that brave, still undaunted band at Harrison in 1846.

"It was believed by the Scranton Company [no longer Scrantons, Grant & Co.] that good 'T' rails could be furnished the Erie Railroad Company, especially on the Delaware and Susquehanna Divisions, on terms more advantageous to the interests of the road then it had before enjoyed."

The race was on! The Scrantons were offering to deliver the new style railroad rail, the "T," under "a contract with the Erie Railroad Company for 12,000 tons of iron rail; to weigh fifty-eight pounds to the yard; to be made and delivered at the mouth of the Lackawaxen [River], in Pike County, during the years of 1847-8, at $70 per ton," ten dollars per ton less than the Erie was paying to bring in rails from England.

No one else in America had ever made that much "T" rail nor at that price. "To the Scranton Company, it was everything!"

Now, the firm would be Scrantons & Platt, starting on November 7, 1846 - this time with three Scrantons, George, Selden, and Joseph, and Joseph Platt as general partners. (Both George and Selden opted to live at Harrison.) Added to them were Philip Mattes, Edward Mowry and John Howland as special partners. And added to them were William E. Dodge, Anson G. Phelps, Benjamin Loder, Samuel Marsh, Henry Shelden, John I. Blair, James Blair, William B. Skidmore, James Stokes, Philip Dater, Daniel S. Miller, John A. Robinson, William Henry Shelden, and Frederick Griffin, more special partners.

Some of these special partners would bear names familiar to students of American history; some of these names would belong to men who would come to live in Scranton, as the community, by common usage, would eventually be called.

"On the 23rd of July, 1847 the first [stationary] steam engine between Carbondale and Wilkes-Barre was started and two rails made.

"*On August 9th, 1847, the mill commenced turning out rails regularly for the Erie Company, which from that date were shipped in every possible way, the roads in all directions leading toward the railroad being full of teams hauling rails or returning with empty wagons. Some days, more than seventy loads were sent off.*"

The race was won!

The race had been against time. The race had been against failure. The race had been against a loss of $3,000,000.

By delivering by teams "some seven thousand tons of rail, which they [Scrantons] delivered at Narrowsburgh, Cochecton, Equinunk, Stockport, Summit, and Lanesboro, an average distance of about fifty miles, thus enabling the [Erie] company to lay track almost simultaneously at all points along the Delaware division as fast as the grading was ready, and open the road for one hundred and thirty miles four days ahead of the appointed time," the Erie would not be in default on its contract with New York State. The Erie had reached Binghamton, New York, on December 28, 1848, before the start of 1849. It would not have to return payment on its multi-million dollar loan from New York.

"The difficulty of carting so large an amount of iron in so short a time can be inferred by those familiar with the mountainous roads and the wilderness intervening."

That little band of men at the confluence of the Lackawanna River and Roaring Brook had at last succeeded.

The Scrantons continued to prosper, and on March 10, 1853, with additional investors, William Henry's vision became the Lackawanna Iron and Coal Company.

George Scranton, as guardian of the company's future, also saw to it that the other part of Henry's vision was carried out.

As the iron "T" rails were pouring out of Scranton in 1847, "The books of subscription were opened at Kresler's Hotel" as the first step towards "completion of a *locomotive engine road* from Great Bend to the Delaware Water Gap," the future Delaware Lackawanna and Western Railroad!

The once forgotten land along the banks of the Lackawanna River would, by 1856, have its own name - Scranton, have its own iron works, its own supply of anthracite, and its own transportation to take its goods to market. But unlike all its predecessors, the transportation would not be by canal, but by railroad.

Chapter 35

Before the Industrial Revolution, man knew only one economic principle - that the man who held land, held wealth.

For those of us born into freedom, there can never be total comprehension of what it is like to live in a world that has just emerged from under the yoke of a monocratic political system.

For those born in this country immediately after the American Revolution, the shackles of a feudal agrarian society based on lord and serf relationship did not slip immediately into oblivion, as if they had never existed.

For those originally born into an agrarian, monocratic political system who can succeed in breaking free, a tiny, partial segment of the shadow that was their past will never leave them.

The American colonists responsible for the American Revolution could never totally shake off all the old habits and way of life. The American officers who led the War of 1776 to victory could never shake off all the old habits and way of life.

Even though Washington, and all the others who had wanted a United States of America, watched carefully in the days following the American Revolution to assure the classless society that we know today as the democratic United States of America, there was, nevertheless, that shadow trace, a hold-over in people's thinking, that would not be eradicated.

The Society of the Cincinnati is one example. The very men who led the way into battle, who as officers risked the charge of treason should they fail in their war for freedom and equality, could not totally surrender the concept of hereditary power and wealth.

They formed their society, open only to American officers, and provided that on their death, membership could be extended only to their oldest sons, a self-perpetuating group of the new country's first leaders.

Washington had seen where the Society could turn into an attempt at establishing a new aristocracy in place of the old, existing one of England - he would not participate, even though the Cincinnati automatically granted him membership. (The Society had made for Washington a special set of china with the insignia of the Cincinnati on each plate.)

The old habits and way of life that were imbedded the deepest was the relationship between man and the land.

Before the Industrial Revolution, man knew only one economic principle - that the man who held land, held wealth.

While the Industrial Revolution was still on the horizon, in the 1780s and 1790s, no one had as yet fully comprehended how and what it would affect. There was as yet no base for wealth by industry alone.

That shadow trace, then, that was man's frame of reference for living, that he would carry with him into another age, was the economics of an agrarian society.

The only change acknowledged as taking place due to the American Revolution at that time was the assurance that in a free, democratic society, the man who held land need not fear that "his lord" could capriciously take it from him to give to somebody else.

The Bill of Rights and Constitution were used to specifically describe just how few the circumstances were under which man need fear such a loss.

And if his government took his land from him, he knew he would be compensated.

(This held true even for states. When Connecticut lost its title over northeastern Pennsylvania by the Decree of Trenton in December 1782, compensation came to Connecticut from the new federal government in the form of the Western Reserve, over 500,000 acres of land in Ohio.)

The man who held land, held wealth.

This was the traditional view. Traditional views will change, but slowly, even in the minds of the people who put the change into effect.

In the great, vast country of America, large landholdings would be anticipated; it would not be an anomaly.

What was unusual was the fact that anyone, if he had money, could acquire acres and acres of land. Land was no longer a prize to be bestowed by the reigning monarch on a favorite. The concept that land is never sold, but passed by inheritance to the next generation, was wiped out by the American Revolution; but the concept that land is the base of all economic value remained intact.

The shadowy trace of the past, then, was the concept that land was wealth and wealth meant large landholdings, and that there could be no wealth but by land.

So all across the new country, new American citizens reached out - to own land.

The men who joined with John Nicholson in 1789 to form The Society for Promoting the Improvement of Roads and Inland Navigation (in Pennsylvania) had all been born before July 4, 1776. They understood the concept of land ownership; it was ingrained in them. They were all owners of large tracts of land throughout southern and western Pennsylvania as well as in the northeastern part of the state.

In fact, that had been one of their motives in forming the Society - to act as a PAC (political action committee) that would pressure Pennsylvania Legislators into appropriating funds for building roads and transportation systems to their lands. The organizers of the Society were John Nicholson, Robert Morris, William Bingham, John Adlum, James Wilson, Henry Drinker, and Thomas Mifflin.

James Wilson, a signer of the Declaration of Independence and a U.S. Supreme Court associate justice starting in 1789, acquired the 12,150 acre Wallenpaupack Manor property from William Penn's heirs in 1793. By the time of his death in 1798, Wilson "owned more land than anyone else in Wayne County." This was in addition to the one million acres he owned in the western part of Pennsylvania.

Wilson was perhaps the most enlightened of all the major land owners who came into northeastern Pennsylvania. He seems to have favored immediate commercial development in contradistinction to individual lot sales spread over a long period of time.

Today, Wilsonville, near the Lake Wallenpaupack dam, is the only reminder of Wilson's attempt to establish a flax (linen) and hemp factory. And the attraction of Nicholson's Union Society (see below) as an investment for Wilson lay in its purpose - 3,000 acres in Wayne County devoted to the manufacture of sugar, potash and pearl ash.

Members of the Drinker family, Philadelphia Quakers, whether considered collectively or individually, constituted another great landholder in northeastern Pennsylvania.

The Drinkers, like Wilson, believed in giving a purpose to whatever community was rising on their land. The senior Henry Drinker resorted to a masterful stroke of publicity when he obtained the endorsement of George Washington - the President of the United States of America! - to help sell lots in Wayne, now Susquehanna County, in the 1790s:

"New York, June 18, 1790

"SIR [Henry Drinker]: - Mr. [Robert] Morris has presented me, in your name, with a box of maple-sugar, which I am much pleased to find of so good a quality. I request you to accept my thanks for this mark of attention; and being persuaded that considerable benefit may be derived to our country, from a due prosecution of this promising object of industry, I wish every success to its cultivation, which the persons concerned in it can themselves desire. I am, Sir, your most obedient servant, George Washington."

The "ad" apparently worked very well, because two years later, on September 18, 1792, a formal name was given to the land development - Union Society; and the community-to-be was officially dedicated to "promoting the manufacture of sugar from the maple-tree and furthering the interests of agriculture in Pennsylvania."

The most well known of all the large landholders was Matthias Hollenback, of Wilkes-Barre. Hollenback's interest began with his arrival in the Wyoming Valley in 1771. He held land titles under the Connecticut claim, and extended his merchant/trading posts into New York State, across the Pennsylvania border. At the time of his death in 1829, Hollenback's land interests totaled almost 400,000 acres.

At the core of the success of the Virginian-turned-Yankee trader was his instinctive ability to respond to the needs of the people around him. At the time of the Wyoming Massacre, Hollenback returned to the fleeing settlers, his friends and neighbors, after riding to inform Captain Spalding, camped at Merwine's (in today's Monroe County), of the events of the day. His reason for returning was to bring bread to the evacuees as they began their long trek back to Connecticut.

Hollenback concentrated his mercantile activities along the path of the Susquehanna River. He usually set up a farming operation along with a store/trading post. His land was for sale, but he was not an active colonizer like the Drinkers and Wilson.

The lands that John Nicholson and Robert Morris accumulated in the 1780s and 1790s became a problem as the two real estate developers and financial geniuses slipped towards debtor's prison. Buyers in America and throughout Europe were hastily sought and courted, in some instances with success.

By an associate, James Greenleaf, Nicholson and Morris were able to tap into wealth held by Dutch financiers; and Theophilus Cazenove, the Holland Land Company's American agent, worked closely with Nicholson and Morris. But for all the activity, it was still not enough to keep the Nicholson and Morris landholdings intact.

Nicholson died in 1800 and Morris in 1806. Their estates passed to their heirs but were so encumbered that the Commonwealth of Pennsylvania, having placed a lien on more than four million acres, established a separate Nicholson Court and Commission to adjudicate claims.

The Court was abolished in 1842. The Morris estate, however, was not settled until the 1880s.

Over 40,000 settlers were recognized as holding their land titles through Nicholson, Morris, or one of their companies. So for all their convoluted land crises, the two men achieved some of their openly avowed goals - to help in the settlement and development of this country.

In one of their desperate efforts to raise money, Nicholson and Morris found a major buyer in the Philadelphians Samuel Meredith and his brother-in-law George Clymer.

Meredith's father, Reese Meredith, had given the equivalent of $25,000 in silver for food and clothing when the Continental Army was at Valley Forge in the winter of 1777. Later, in 1780, the son and Clymer would also pledge $25,000 for more supplies for the American army. Clymer affixed his name to the Declaration of Independence that famous July 4, 1776; and Samuel Meredith served as treasurer of the United States for 12 years, under George Washington and John Adams.

The two brothers-in-law were already large landholders when they acquired some of the Morris/Nicholson tracts. By 1800, Samuel Meredith and George Clymer had come to stand in the place of Nicholson and Morris as the great landholders of northeastern Pennsylvania. All told, they had almost two million acres spread over northeastern Pennsylvania, New York State in Delaware and Sullivan Counties, western Virginia, and Kentucky.

Shortly after 1802, Meredith decided to live in Wayne County, at Belmont in Mount Pleasant Township. Some of his children also settled in northeastern Pennsylvania and became active in community affairs.

In addition to Nicholson and Morris, William Bingham was the only other member of the Society for Improvement to own many acres of land throughout the United States. He was a Philadelphia financier, who at one point served as American commissioner to Martinique. As early as 1786, Bingham began purchasing Maine timberland from Massachusetts. He eventually acquired two million acres in Maine alone. Binghamton, New York, is a reminder of another tract of land held by the financier. Much of Bingham's land passed to his heirs.

Chapter 36

The significance of such strong Quaker ties in the earliest days of Providence/Scranton's development is that it set, unconsciously, the character of the community.

Members of the Tripp and Slocum families who settled in Providence Township (now Scranton) could not be considered large landholders in the same sense as Hollenback, Drinker, Wilson, Bingham, Meredith and Clymer, etc. However, their land ownership was significant because it was concentrated in one area. Almost all of Providence Township was their property. It was Providence Township that later became the City of Scranton.

About 1800, Stephen Griffin arrived from the Oblong area in Dutchess County, New York. He was soon joined by brothers Joseph, Thomas, and James. Like the Tripps and Slocums, the Griffins were Quakers. By 1830, the Griffins acquired over 40% of what had once been Tripp owned lands. This was in addition to what they had purchased from other landowners. When taken collectively, Tripps, Slocums and Griffins claimed 10,200 acres of the 16,083 acres that made up Providence Township.

(New York's Oblong was an area along the eastern border between New York and Connecticut. Sparsely settled, Quakers, starting about 1740, moved into the strip and established a series of Meetings. "Each Quaker home as it settled became a resting place for those who followed, for it was a cardinal principle of Quaker hospitality to keep open house for all fellow members, under all circumstances.")

The three Quaker families, Griffin, Tripp, and Slocum, devoted much of their energies to opening up, clearing, and settling Providence Township.

At one point, Stephen Griffin blasted rocks near Babylon, about a half-mile below Old Forge on the Lackawanna River, so that he could send timber to the Susquehanna River and to market. For a while the area he had cleared in the Lackawanna valley was known as "Quaker Falls."

When Stephen Griffin first left for Providence, he and his wife Mary (Place), with some of their children, obtained permission to leave the Society of Friends Stanford Monthly Meeting, in Dutchess County, New York, for the Meeting (unnamed) at Providence, in Luzerne County, Pennsylvania.

But by 1815, Stanford Meeting reported that Stephen and Thomas Griffin, "having some time since removed with [their families] and settled in the town of Providence and Luzerne County," had their certificates of membership forwarded to Muncy Monthly Meeting (about 70 miles west, near Williamsport).

In 1817, two years later, the Stephen Griffins were accepted as members at Roaring Creek Monthly Meeting, a short distance southeast of Muncy.

But by 1825, Stephen's other brother James Griffin, with his wife Sarah (Clapp) were certified to leave from another Oblong area Meeting, Nine Partners, for the Meeting at Stroudsburg, Pennsylvania, as they would be living in Providence, Luzerne County.

So, the Meeting at Providence never materialized; and those who did not want to openly break with the faith of their forefathers attached themselves to a far off Meeting, too far to attend with any regularity.

The significance of such strong Quaker ties in the earliest days of Providence/Scranton's development is that it set, unconsciously, the character of the community.

The Slocums, Benjamin and Ebenezer, the Tripps, Isaac and Job, and the Griffins, Stephen, Thomas, Joseph, and James, were all New England Quakers. Their families had never experienced the protective, privileged community of life under William Penn.

To the contrary, being a Quaker in the New World before Penn's arrival could mean not only harm to the self but to those associated with the Quaker. The Dutch New York government, as early as 1657, legislated a fine of £50 for anyone caught harboring a Quaker; and any vessel bringing Quakers to New York City could be confiscated.

So, there would be an air of inherent reticence to the little community midway along the length of the Lackawanna River. The preference for dull coloring, like Quaker dress, carried over to houses and buildings; the plain, the simple was always favored.

But just as there would be no display of worldly ostentation, there would be the belief in freedom - freedom of conscience, freedom of the individual, freedom of expression.

There was yet another aspect to a Yankee-turned-Pennsylvania Quaker, making him even more distinct from his New England and Philadelphia/Penn contemporaries. He had to survive in an isolated wilderness, alone. Many of the tenets and principles of his forefathers would have to fall by the wayside in the fight for survival.

Quakers along the Oblong would be disowned "for complying with military requirements and attending places of diversion." At Wilkes-Barre, in Providence Township, and throughout Wyoming, military expertise meant that the community might continue to exist. And the Slocums, with their forge, their grist mill, and - distillery?

The large landholders of Providence/Scranton, united first by religion, then by intermarriage, would form a closed community that would open to receive like-minded individuals. Later, children and grandchildren would be active in other religious denominations. But the basic personality of the area had already been shaped so that the power and wealth from anthracite would follow a unique path. These were the people who would be William Henry's and the Scranton brothers' neighbors when they purchased the land at Slocum Hollow.

Chapter 37

No doubt about it, the activities of Pulteney Associates would be of importance in northeastern Pennsylvania.

Meanwhile, north of Pennsylvania a scenario akin to the Connecticut/Pennsylvania jurisdiction claim was being acted out by Massachusetts and New York.

On May 27, 1784, a year-and-a-half after the Decree of Trenton had settled the question of jurisdiction over the northern one-third of Pennsylvania, Massachusetts petitioned the U.S. Congress to settle the dispute between itself and New York. Massachusetts was claiming all of New York State from the Owego River west to the Niagara River. When Congress failed to act, the two states involved met and worked out a settlement on December 16, 1786.

According to the terms of the agreement, New York would have jurisdiction over the contested lands and Massachusetts would have right-of-soil. The amount of land involved was about seven million acres.

Massachusetts obtained Indian title from the Senecas to 240,000 acres between the Owego and Chenango Rivers, which were immediately sold to a Boston company, becoming known as the Boston ten townships. The remainder of the land was bounded on the east by a line running north from the 82nd milestone on Pennsylvania's border to Lake Ontario; on the south by the Pennsylvania border; on the west by a jagged line set one mile east of the Niagara River; and on the north by Lake Ontario. (Mansfield and Mansfield University, in Tioga County, Pennsylvania, are about six miles west and 12 miles south of Milestone 82.)

It took another year and a half (until April 1, 1788) for Massachusetts to find a buyer, New Englanders Nathaniel Gorham and Oliver Phelps; the almost seven million acres would be sold for £300,000 in Massachusetts currency, payable in three installments. But there was no Indian title to this acreage, so Phelps and Gorham also had to purchase from the Senecas, which they did July 8, 1788. The Senecas, however, would only sell the eastern portion, about 2,200,000 acres, in this particular sale. Afterwards, the land became known as the Genesee Tract.

The new entrants into the world of land development were now on a par with John Nicholson, Robert Morris and William Bingham. They opened a land office at Canandaigua (Ontario County) and soon succeeded in selling about two townships of land.

But when the annual installment of their payment to Massachusetts came due in 1790, Phelps and Gorham had difficulty. A hasty note went out to Robert Morris, and on November 18, 1790, for $5, Morris became the sole owner of all of the remaining two million acres of the eastern Genesee Tract, and with a claim on the remaining western portion.

Morris worked immediately on converting as much of his newly acquired real estate into cash as possible. The first buyer to come forward came through Morris' London office. Three gentlemen organized a syndicate and paid $333,333.33. All three men were British citizens. Sir William Pulteney, who held nine-twelfths of the syndicate shares, was an English capitalist. William Hornby, with two-twelfths of the syndicate, had recently retired as governor of Bombay, India. Patrick Colquhoun, a Glasgow, Scotland, lawyer, held a one-twelfth share in the syndicate. The syndicate became known as the Pulteney Associates.

Everything looked rosy and fine. Phelps and Gorham were out from under an embarrassing debt. Morris had raised the necessary money. And new land speculators had made a purchase worthy of their talents. (Through Theophilus Cazenove, Morris sold his interest in the western portion to a group of Dutch financiers and title became vested in The Holland Company.)

But when man proposes, government imposes. Prior to 1800, New York law would not allow foreign nationals to hold title to property located within New York State.

So much for the syndicate.

"It was therefore necessary that the syndicate should select a person who could assume the title and convey such lands as they deemed it advisable to sell. Captain Charles Williamson was chosen..."

When Williamson with his family arrived in Philadelphia, he met with Morris. On January 9, 1792, he became a naturalized U.S. citizen. Williamson then proceeded to inspect the Genesee Tract, traveling from Philadelphia to New York City, then to Albany, and southwest from there into the Genesee. On returning to Philadelphia on April 11, 1792, Williamson acknowledged the transaction for the syndicate and received a deed from Robert Morris.

Consulting with Morris for a more direct route to the Genesee, Williamson established his family at Northumberland in Northumberland County, at the junction of the East and West Branches of the Susquehanna River.

He then began construction of a road northward, past Williamsport, following the Tioga River from its headwaters to its junction with the Chemung River, and from there northward up the Cohocton River to an open plain he named Bath.

Williamson built a town at Bath, and advertised land for sale from Virginia to Canada. He was a free spender (of the syndicate's money), operating the development on the "hot house" method. Williamson's technique was to clear about 500 acres and build a saw mill, grist mill, and one or two buildings to make it visually attractive to prospective buyers.

It worked. The Pulteney Associates' agent told Morris that by the end of 1794 he had been able to sell 463,017 acres for a total of $657,965. Williamson expected that the Genesee Tract in its entirety would bring over $3,157,965. (Morris was so impressed by Williamson's results that he adopted the same procedures for selling some of his own land.)

But however man proposes, government will forever impose. New York State announced that for a three year period beginning in 1801, foreign nationals could hold title to New York lands.

The syndicate opted for direct title. "Accordingly, there was an accounting and an appraisal; Williamson assigned to his principals $551,699.78 worth of bonds, mortgages and notes, and on March 31, 1801, conveyed to them the unsold lands which were valued at $3,547,494.58; besides the original purchase, he conveyed over 5,000 acres of land just west of the Genesee, the Cottinger tract in the Morris Reserve, 600 acres in the Military Tract, besides thousands of acres in Otsego, Herkimer, Chenango, Clinton, Albany and Montgomery Counties. Under his administration he had expended in purchasing lands, making improvements and other expenses, $1,374,470.10; he had received for lands sold, $147,974.83; besides this there was an indebtedness outstanding of about $300,000, the most of which was the unpaid purchase price of land outside of the original purchase."

Included in Williamson's assignment of lands to Pulteney Associates was 11,526 acres between the Lackawaxen and Big Middle Creeks in Wayne County. The tract had been a part of Robert Morris' holdings in Wayne County in the 1780s. No doubt about it, the activities of Pulteney Associates would be of importance in northeastern Pennsylvania. As soon as he had rendered his accounting to his employers, Williamson resigned. Robert Troup was appointed general agent, Pulteney Associates.

Chapter 38

When Troup died in 1832, Joseph Fellows was named by the Pulteney Estate to replace Troup as its general agent. The power and prestige, and the responsibilities that had been Troup's now belonged to Joseph Fellows, Jr.

The Joseph Fellows family was among the new arrivals at Providence Township (Scranton) in September of 1795. Family members included the father, Joseph, his wife Catherine (Turvey) and seven children. Of their children, Joseph Fellows, Jr. was the oldest, born at Redditch, Worcestershire, England on July 2, 1782.

The following year, 1796, Joseph, Jr. was apprenticed to Isaac I. Kipp, a lawyer in the northern portion of Philadelphia. Joseph was a bright boy, did well under Kipp, and on becoming a lawyer, looked for work; he did not attempt to go into private law practice.

When Joseph found work, it was in New York State under Robert Troup, to attend to legal matters in the Pulteney land office. In 1810, Troup sent Fellows to Geneva, New York, at the northern tip of Seneca Lake, to be a Pulteney sub-agent. Fellows was then

28 years old. His association with the Pulteney interests would continue for the rest of his life, until two years before his death in 1873 at the age of 91.

Shortly after Fellows arrived at Geneva, Troup conveyed Pulteney's 11,000 acre tract in Wayne County, Pennsylvania, to Milford's first attorney, Dan Dimmick. Within a year and half, in January 1817, Fellows had found a buyer for Dimmick's land, George Goundry, of Geneva, New York.

Fellows, however, never severed his ties with his family back in Providence Township (now Scranton), Pennsylvania. To the contrary, he depended more and more on their participation with him as he moved upward in the trust and confidence of Troup and Pulteney Associates.

In 1808, Joseph's younger sister Mary married Philip Heermans. In 1795, the Heermans had settled in Pittston (north of Wilkes-Barre along the Susquehanna River), coming from Dutchess County, New York.

Then there was Henry Heermans, Philip's younger brother...

Henry could understand his sister-in-law's brother. Of all of Joseph Fellows' family, it would be the in-laws who would work the closest with him, Henry becoming the first to do so.

At the time Troup took over management of Pulteney interests in 1801, the drive in New York State for a canal to open up a pathway to the Great Lakes and western United States was of prime interest.

Troup's support was needed to build a canal across New York, as was support from all of the major land holders between the Hudson and Niagara Rivers. For a while, New York officials worked with Joseph Ellicott, manager for the Holland Land Company. The Holland Company gave 100,632 acres in 1814. From William Hornby's agent, John Grieg, another 3,500 acres was given towards a canal.

Troup, on the other hand, held back, waiting to see how things developed, claiming he had not yet received full authority to commit the Pulteney interests.

A year later, the land grants were canceled; for a moment, it seemed there would be no Erie Canal.

(The Holland Land Company would later give 100,000 acres. Company lands were that portion of New York from west of the Pulteney property to the Ohio boundary.)

There would always be an interest in a canal, however. But while canal advocates needed land, and as cheaply as possible, the fact could not be overlooked that Pulteney, Holland Land Company, and other large landowners would derive enormous benefits towards opening up their lands for settlement if a transportation canal existed which traversed their lands.

Troup remained aloof from involvement on the issue of canal land donations because he had already made his move. The Pulteney agent had placed himself with the men pushing for a canal. He knew that from that position he couldn't lose because he was operating "both sides of the coin." He would be one of the ones asking for land donations and he would be one of the ones asked.

Troup joined DeWitt Clinton, Elkanah Watson, Thomas Eddy, Jeremiah and Stephen Van Rensselaer, Gouverneur Morris, Benjamin Wright, Gideon Granger, Jonas Platt, and Simeon De Witt on the board of a company that had been incorporated in 1792, the Western Inland Lock Navigation Company. The Company had already built portions of a canal between Albany and Seneca Lake by 1800. But with the growing demand for a much larger canal, New York State had taken over Western Inland in 1812, making it a part of the Erie Canal project. The same men, now including Troup, continued their leadership under New York State.

Troup's importance can not be underestimated. When the Erie Canal was finally built, from 1817 to 1825, almost 45 miles would be through Pulteney lands.

Troup was considered to be one of the most prominent men in western New York.

Troup's power lay in his control of approximately two million acres. Originally, all of the Pulteney Associates' land was in Ontario County. Then, in 1796, the southern half of the area had been separated off to form Steuben County. By 1822, when the Erie Canal was half completed, the population of the truncated Ontario County was listed at 61,185. And Steuben had climbed to a population of over 22,000.

So not only did Troup represent millions of acres of land, but also had contact with almost 100,000 people, many still owing money to Pulteney on their land purchases.

And regardless of political jurisdiction or county name, the agent for the English syndicate would continue to develop the area, increasing the population far into the future. Even those changes brought about by the passing of time would not interrupt the activities of what had started out as the Pulteney Associates.

As the original shareholders in the syndicate passed away, Troup adapted, becoming agent for the Pulteney Heirs; later, he was agent for the Pulteney Estate. (Hornby Heirs had established their own agent, John Grieg.)

"Bath [in Steuben County] sat among the hills in quiet self-possession and stateliness. There, too, was the land proprietor's representative. About the stiff rooms of the Land Office there was a seclusive chilliness that spoke of English aristocracy and a plethora of capital that must seek relief by investment in countries far remote."

When Troup died in 1832, Joseph Fellows was named by the Pulteney Estate to replace Troup as its general agent. The power and prestige, and the responsibilities that had been Troup's now belonged to Joseph Fellows, Jr.

Thirty years after his death, Fellows was described by one who had known him, "He was always honorable, exactly just and fair in his dealings. He conducted the business of the Estate with strict integrity, and was really, as I believe, as indulgent and liberal as his predecessors had been. There are many of the old settlers who could testify to the kindness he had shown them. Regarded by many as miserly, he was, on the contrary, really a man of charitable impulses and of a good heart."

Chapter 39

Because Blossburg coal was bituminous, was like English coal, it had instant value.

Before Troup's death in 1832, Fellows had begun investing in land in his own right. For $20,000, Fellows found that George Goundry would be willing to part with the 9,271 acres that remained of the tract in Wayne County, Pennsylvania, that he had purchased ten years earlier.

This was at a time when there was a lot of excitement in northeastern Pennsylvania. The Wurts brothers had called on Benjamin Wright, John Jervis, and James Archbald, engineers of the Erie Canal, to come and help them with their project - the Delaware and Hudson Canal.

Fellows knew first hand what a canal could mean to a community. His work under Robert Troup had been a good training period. For his family, living in Providence Township where the chief source of income was from the barter system, a canal could be the means of bringing about a dramatic change in lifestyle.

Fellows first approached Goundry about timber rights for the winter of 1826-1827. When the fellow Geneva resident agreed, Fellows sent out a call to his family. The brother-in-law Henry Heermans was the first in Fellows' family to come and work with him.

When Henry's brother Philip had married Mary Fellows in 1808, Henry had left the family home and gone - not west, but east - to the intersection of the north/south road with old Minisink Path. The place was known as Salem Corners (*today*, it is Hamlin, in Wayne County). There, at the crossroads, lived a cousin on his mother's (Griffin) side, Isaac Dolph. (Before 1800, Dolph had come to "The Corners," where he operated a store and sold land to newcomers. He had originally settled, and still owned, a small amount of land in Providence Township.)

While living in Salem, Henry Heermans worked first as a carpenter; he eventually took over Dolph's store, and married the daughter of a neighbor, Fandini Nicholson. Fanny had nine brothers and sisters.

Among all the Fellows, Heermans, Dolphs, and Nicholsons, there was no one who could match Joseph Fellows, Jr. and what he was doing.

Family members already had proof of that. In 1824, Joseph, Jr. had saved enough money to buy Philip Heermans' interest in 100 acres in Providence for $1,000 cash!

When Joseph, Jr. let it be known he wanted help in developing the Wayne County land, Henry and one of Fanny's brothers, Zenas Nicholson, responded to his call. They began timbering the tract. Later, they built a saw mill near the confluence of the Wallenpaupack and Lackawaxen Rivers (Hawley). Timber sales were immediate - to the D&H Canal Co.

On April 12, 1827, Heermans handed over $4,400 (in cash) to Michael McKeel for 300 acres in Providence Township. On November 5 of the same year, Fellows bought Goundry's 9,271 acres outright.

By 1828, Heermans' Salem Corners store had been sold to Oliver Hamlin and the sawmill to Fuller and Company. Fanny, Henry and their children leaving Wayne County for Main Avenue, Providence Township, Luzerne County - the first of many to leave for Providence.

And Fellows, guided by Pulteney procedures, began selling off parcels of his Wayne County land.

After Troup's death in 1832, more and more opportunities opened up for Fellows. The Erie Canal was just starting to show the benefits its advocates had believed in. Land sales swelled all along its route. By the time of his death, it was estimated that Fellows had signed more deeds than anyone else in all of New York State.

Then in 1835 Fellows formed The Corning Company. Fellows had land to sell; The Corning Company would buy and develop land.

There were nine members of The Corning Company: Erastus Corning, Thomas Olcott, and Watts Sherman, from Albany, New York; Ansel Bascom and Bowen Whiting, from Seneca Falls and Geneva, New York; William A. Bradley, from Washington, D.C.; Levin I. Gillis, from Montgomery County, Maryland; Hiram Bostwick, from Otsego County, New York; and Fellows, still living at Geneva. The company was to exist for 20 years.

Hiram Bostwick, trained as a lawyer, was the most visible of the company members. It was he who had gone to Albany for Fellows and spoken to Corning and Olcott. Corning was head of Albany City Bank; Watts Sherman was a cashier in Corning's bank. Olcott was head of the Mechanics and Farmers Bank, also located in Albany.

But there was more to The Corning Company than just buying land from the Pulteney Estate for development. Corning later wrote that they were "engaged in the building" of a railroad.

When Captain Charles Williamson arrived in Philadelphia in 1792 to begin his duties as Pulteney agent, he found that his most immediate problem was the need of a road to access the property.

Williamson finally settled on approaching the Genessee Tract from the south, cutting a path almost due north from the juncture of the East and West Branches of the Susquehanna River at Northumberland, Pennsylvania.

In going through what is now the southern portion of Tioga County, Williamson's men showed him a strata of black mineral where they had cut into some rock. It was not the unusual "black stones that burn." It was, instead, a black mineral the British agent was familiar with - soft, bituminous coal.

For Williamson, the knowledge of the secret nestled deep in the Endless Mountains was his. But the practicality of what to do about it was beyond him at that point in time. (Eventually, two settlers, Aaron Bloss and David Clemons, would mine and ship small amounts from their Tioga County farms.)

Over time, however, more and more people became aware of Tioga County coal. Starting in 1806, Aaron Bloss built and operated a hotel at what had once been Peter's Camp on Williamson's road. Today, it is known as Blossburg.

Bloss began using the coal he found just for himself. But gradually, travelers stopping at his hotel spread the word, until finally Bloss was shipping small amounts south to Philadelphia and north into New York.

In 1841, Sir Charles Lyell, "the celebrated Scottish geologist and close friend of evolutionist Charles Darwin," visited Blossburg. When he returned to England, he published:

"It was the first time I had seen the true coal in America, and I was very much struck with its surprising analogy in mineral and fossil character to that of Europe; the same white grits or sandstones as are used for building near Edinburgh or Newcastle; similar black slates, often bituminous, with the leaves of ferns spread out as in an herbarium, the species being for the most part identical with the British fossil plants; seams of good bituminous coal, some a few inches, others several feet thick; beds and nodules of clay, ironstone, and the whole series resting on a coarse grit and conglomerate, containing quartz pebbles very like our (British) millstone grit, and often called by the American as well as English miners 'farewell rock,' because when they have reached it in their borings they take leave of all valuable fuel.

"Beneath this grit are those red and gray sandstones corresponding in mineral character, fossils and positions with our old red.

"I was desirous of ascertaining whether a generalization recently made by Mr. Logan in South Wales could hold good in this country [United States]. Each of the Welsh seams of coal - more than ninety in number - have been found to rest on a sandy clay or firestone, in which a peculiar species of plant called *Stigmaria* abounds to the exclusion of all others.

"I saw the *Stigmaria* at Blossburg in abundance, in heaps of rubbish extracted from a horizontal seam.

"Dr. Saynisch, the president of the mine, kindly lighted up the gallery that I might inspect the works, and we saw the black shales in the roof adorned with beautiful fern leaves, while the floor consisted of an under clay, in which the stems of *Stigmaria*, with their leaves and rootlets were attached, were running in all directions.

"The agreement of these phenomena with those of the Welsh coal measures, 3,000 miles distant, surprised me, and led me to conclusions respecting the origin of coal from plants not drifted but growing on the spot."

Because Blossburg coal was bituminous, was like English coal, it had instant value.

Chapter 40

A charmed circle grew up around Joseph Fellows - sometimes the landowner would also participate in activities associated with the use and development of that land.

Long before Sir Charles Lyell visited the mine at Blossburg, Pennsylvania, New York and Pennsylvania Legislators had been pressured to provide a way of accessing the coal.

The Chemung Canal, from Seneca Lake to Elmira, New York, was completed in 1833. Navigation on the Pennsylvania side of the Tioga River was to be improved by The Tioga River Navigation Company so coal, iron, and timber could be sent north to connect with the Chemung Canal.

As the Chemung opened, in 1834 Philadelphians headed by James R. Wilson, Charles F. Seidel, Joseph P. Morris, M.D., with Lewis Saynisch, M.D. of Blossburg, purchased approximately 7,000 acres in the Blossburg area of Tioga County. The men would mine the semi-bituminous Blossburg coal for shipment into New York State where it was guaranteed steady sales.

The white gold that was the salt mines at Syracuse, New York, needed all the Blossburg coal it could get in order to process the salt.

But even though the Chemung Canal would eventually be extended to Corning, the over-all water route from Blossburg was too difficult, and was finally abandoned. In its place, a railroad would be built to join the Chemung Canal.

The Tioga Navigation Company assumed responsibility for constructing the Pennsylvania portion of the railroad. Tioga Coal, Iron Mining and Manufacturing Company, of New York, would be responsible for the New York portion of the railroad.

Meanwhile, The Corning Company, after it was formed in 1835, began to lay out the city of Corning, New York. Corning would be the terminus for the railroad coming up from Pennsylvania, the railroads meeting at Lawrenceville on the Pennsylvania side. The other terminus would be at Blossburg.

But by the beginning of 1838, the railroad had not yet been completed, Tioga Navigation having financial problems, and Tioga Coal and Iron as well. Failure to complete an adequate transportation system was having a disastrous effect on the coal trade, too.

So, on April 13, 1838, Wilson and Saynisch, with James H. Gulick, Franklin R. Smith, M.D., Joseph P. Morris, M.D., William Frederick Seidel, and others, including a member of the Corning Company, Bowen Whiting, joined together to form the Arbon Coal Company. Arbon Coal could own up to 2,000 acres of land.

Then, two weeks later, on April 27, 1838, the Arbon Land Company was formed by just the land owners of 1834, Seidel, Morris, Saynisch, and Wilson, with Frederick A. Lohse concurring. (Lohse had bought some of Saynisch's land interest in the meantime.) These men placed the 7,000 acres they owned at Blossburg and surrounding area in the hands of Franklin R. Smith, M.D., Bowen Whiting and James H. Gulick, to be held in trust.

The trustees were to use the 7,000 acres as backing for 4,000 shares of common stock of Arbon Land Company with a value of $50 per share. Of the 4,000 shares, one-half, or 2,000 shares, was to be divided equally between Saynisch, Seidel, Morris, Wilson, and Lohse.

The other 2,000 shares were to be sold for a total of $100,000. In addition, thc buyers of Arbon Land stock would be "entitled to receive certificates of stock or loan [bonds] from either of the railroad companies" (Tioga Navigation or Tioga Coal and Iron).

Seven months later, over in Corning, The Corning Company established its member, Hiram Bostwick, as president of the bank it formed, the Bank of Corning. In 1841, Arbon Coal Company mortgaged its Tioga County land, almost 2,000 acres purchased from the trustees of Arbon Land Co., to the Bank of Corning.

The trust that had been set up under Arbon Land Company continued for many years. The trustees actively engaged in buying and selling: additional lands were purchased; lots were sold for building houses, schools, etc.; and as mining continued, new areas near Blossburg opened up, such as Arnot, Fall Brook and Morris Run.

The beneficiaries of the trust were the shareholders of Arbon Land Company stock. By 1858, there had been some changes in stock ownership. Gulick owned 55%, over one half. Corning Company members William A. Bradley, Erastus Corning, and Thomas Olcott each held about 2%. Their investment was in sharp contrast to another member, Joseph Fellows, whose investment was three times that of any other Corning Company members: Fellows held 6%.

(Another Company member, Bostwick, had held the same 2% as Bradley, Corning, and Olcott in 1848, but by 1858 he was no longer a shareholder.)

The addition of several other New Yorkers as shareholders in Arbon Land, however, indicated that Blossburg coal activities of the 1850s and 1860s were still connected to The Corning Company.

Arbon Land Company shareholders were entitled to percentage interests in the lands and all other assets held by the trustees. The percentage interest was in direct proportion to the number of shares owned by each shareholder. But while the stockholders could claim the land and other assets tied to their company, Arbon Land, it did not mean that they could also automatically claim an interest in the companies or individuals authorized to conduct business activities using the land. (The only exception to this had been the tie in with the railroad in 1838.)

What this would mean in generic terms was that as a landowner, one would derive income/profit by leasing one's property to someone else. The lessee, in turn, would have the right to derive income/profit by using the same piece of land as he saw fit, even to removing parts of the land, such as minerals, coal, sand, etc., if that was the purpose of the lease.

As an alternate way of deriving money from the property, the landowner could make an outright sale of the minerals in and on the land to someone else.

Once the owner has concluded his agreement with the lessee, or mineral buyer, he has received all the profit he is entitled to receive from his land ownership. To receive more benefits from the same piece of land, he would have to join with and become an owner, partner, or shareholder with the "someone" or entity who was leasing his land or buying his mineral rights.

And that was what happened in developing the Blossburg coal fields. Starting with the entry of The Corning Company, some men not only provided funding for their own project but also participated in every project along the way.

A charmed circle grew up around Joseph Fellows - sometimes the landowner would also participate in activities associated with the use and development of that land.

And in that "charmed circle," the landowner, the fee owner, very rarely assumed a highly visible position, such as leader or president of the operating company; at most he might hold the title of company/corporate treasurer.

Arbon Land Company was not the only entity with its assets tied up in a trusteeship. The Corning Company was operating in the same fashion, although its trustees changed from time to time. Ansel Bascom, yet another Company member, was the first trustee to conduct business for The Company. He was joined later by fellow members Whiting and Bostwick. Then, after November 1840, Fellows and Corning served as trustees.

Sometimes, Fellows and those around him were in a situation where at every level of business the day-to-day transactions were being handled through trusts and by trustees. Trust assets could include stocks, bonds, patents, royalty rights, etc. as well as real estate.

The railroad, named Blossburg and Corning R.R., and running between Blossburg and Corning, was completed in September, 1840. This was two and a quarter years after Arbon Land wrote that the land in Tioga County was "abounding in Bituminous Coal, Iron Ore and other valuable minerals," and "to accomplish which it is necessary that a Rail Road should be made therefrom for their transportation."

About the railroad, Anton Hardt recalled in 1909, "I took my first ride over the road [on] January 7, 1867, and remember well the striking appearance of the passenger coach, which was coupled to a long string of five-ton coal dumps. In the center of the coach was a good-sized coal stove, a plush covered sofa on each side of the coach opposite the stove, and seats arranged in the usual way at each end of the coach. There were only two ticket offices on the line, one at Corning and one at Blossburg. It took three and one-half hours to cover forty miles from Corning to Blossburg."

In looking back over 40 years, Hardt also commented, "When we consider the state of the country, the building of the Blossburg and Corning Railroad, in 1840, must be regarded as one of the boldest enterprises of that day," adding,

"The Blossburg and Corning R.R. was, therefore, one of the oldest railroads in the United States, built expressly to carry coal, and, as events proved, it was somewhat in advance of the wants of the country."

Arbon Coal shipped 4,234 tons that year, and a total of almost 50,000 tons by 1845. The bituminous "soft" coal went north not only to the salt at Lake Onondaga, but also north and east to Albany and Troy, New York. There Erastus Corning owned iron works and a small railroad, the Utica and Schenectady. His own companies bought and used Blossburg coal; his own iron works manufactured the rails used in building the Blossburg and Corning R.R.

When Arbon Coal Company ceased its mining operations about 1845, William Mallory & Company, sometimes called Mallory and Bostwick, began mining. In 1859, John Magee and the Fall Brook Coal Co. continued, but slightly east of the first mines at Blossburg. Mallory, Bostwick, and Magee were all residents of Corning, New York. Mallory and Bostwick were shareholders in Arbon Land. Magee, starting up 20 years after the first coal trains left Blossburg, nevertheless had that perennial trustee of Arbon Land Co, James H. Gulick, as one of the three charter members of the Fall Brook Coal Co.

Were all these activities a trust? Was this what the Sherman Antitrust in 1890 and Clayton (1914) Acts were enacted to stop?

Chapter 41

In 1853, Joseph Fellows, Jr. was 71 years old. The men who depended on him, the men who surrounded him, the men who benefited from his actions were all wondering how much more he could do?

Hiram Bostwick was the most visible member of The Corning Company when it started in 1835.

It was Bostwick who had traveled to Albany, New York to contact Erastus Corning and Thomas Olcott; and it was Bostwick, for The Corning Company, who purchased the first land in Steuben County, New York, for the future site of Corning.

Bostwick had grown up in Otsego County, New York, left there to study law at Albany and afterwards to work for Joseph Fellows. There is a silent testimony to Bostwick in the Quaker Burying Ground in Tioga County. At Knoxville, in Deerfield Township near the New York border, a gravestone commemorates the death of a young Hiram L. Bostwick, age 19, who died at the Battle of Bull Run on August 28, 1862.

Members of the Society of Friends were very much against violence of any kind. But at the same time, their belief in the "inner light" that guided an individual's conscience was so strong that conceivably a Quaker could be 'moved' to fight.

Otsego County, where Bostwick grew up, was predominantly Quaker in the first part of the 1800s. *Today*, everyone is familiar with Otsego County because that is where Cooperstown and the Baseball Hall of Fame are located.

But before baseball, the great significance of Otsego County was that the headwaters of the Susquehanna River arose within its boundaries; the water that flowed from Otsego Lake did not stop until it emptied into the Chesapeake Bay.

Four large land owners came into Otsego immediately after the Ft. Stanwix Treaty was signed in 1768. Richard Smith, a Quaker from Burlington, New Jersey (across from Philadelphia), hoped to start a Quaker community on 69,000 acres that he and several other gentlemen were developing. The land was known as the Otego Patent.

Immediately to the west of Otego Patent, John Butler had held comparable acreage. This John Butler was the same Butler who led the British, Indians, and Butler's Rangers against the American Yankees at the Battle of Wyoming on July 4, 1778.

To the north and east of Otego, the owner of 100,000 acres, including all of Otsego Lake, had been Colonel George Croghan, Sir William Johnson's deputy. After Croghan mortgaged and lost his land to William Franklin (Benjamin Franklin's son and a governor of New Jersey), the 100,000 acres passed to Burlington (New Jersey) Quakers William Cooper and Andrew Craig. Cooper and his family made their home at Lake Otsego. His son, James Fenimore Cooper, was to become one of the giants of American literature.

On land about the same size as Smith's and due north of it, John C. Hartwick established another settlement.

By 1790, a Smith son had left his father's area and opened up a store on Cooper's tract.

By 1825, the Quaker way of life was so dominant in Otsego County that legal documents were often recorded at Cooperstown (the county seat) using exclusively Quaker phraseology.

This was Hiram Bostwick's background.

This was also the background of William W. Winton.

This was also the background of Horatio S. Pierce.

When Bostwick left to join Joseph Fellows, Andrew Winton and his son William Winton left for Providence Township/Scranton, as did Levi Pierce, the father of Horatio Pierce, and his family. Horatio, however, did not remain in Providence. He moved on to Carbondale, the terminus of the Delaware and Hudson Canal and Gravity Railroad system.

At Providence, William Winton married Henry Heermans' oldest daughter, Catherine, in 1835. Joseph Fellows' nephew, John Fellows, married Cynthia Jennitte Pierce, Horatio Pierce's sister.

After Henry Heermans moved to Providence Township, he opened a store and began acquiring land where coal was known to exist. His son Alvah married Sarah Slocum; his son Aretus married Mahala Jane Tripp. And daughter Mahala married Abram (A.B.) Dunning; and daughter Polly married Isaac Dean.

Heerman's wife Fanny kept in close contact with her family back at Salem Corners/Hamlin, in Wayne County. Her mother had remarried, to John Weston. Eventually, Weston grandchildren made Providence their home, too.

Some of the children of Fanny's brother, Zenas Nicholson, also made Luzerne County their home: Horatio W. Nicholson became a lawyer and helped in the founding of the Delaware, Lackawanna & Western Rail Road; his half-brother, George Byron Nicholson, became a promiment lawyer in Wilkes-Barre.

After Fanny died in 1831, Henry Heermans remarried. He died in 1843. His second wife, Sara Ann, later remarried, to Joseph Steele, Jr. whom she subsequently divorced. When Heermans died, his will left everything equally to his children, with a life interest to his widow. William Winton was named his executor.

Winton continued on with his father-in-law's store, taking Charles Atwater in as a partner. Later, A.B. Dunning became Winton's partner.

Horatio Pierce, who had gone to Carbondale, began as a clerk in Gillespie & White's store. The D&H was actively mining at Carbondale, and the town was mushrooming like a typical boom town. There was no place north of Wilkes-Barre bringing in money like the D&H at Carbondale and at its canal terminus in Honesdale.

It wasn't long before Pierce became Thomas Gillespie's partner and the firm was known as Gillespie & Pierce.

To the north of Carbondale, in Susquehanna County, two communities had started up - one called Montrose, the other Dundaff.

While a small settlement had existed previously, Dundaff really began in 1822 when Colonel Gould Phinney purchased the land. Later, Phinney left Wilkes-Barre to actively develop his Susquehanna property. At Dundaff, Phinney's daughter Rachel married a neighbor, John Jay Phelps, and they moved to New York City. John Jay's brother Horace G. Phelps moved first to Corning, New York, and then to Scranton, Pennsylvania; brother Jaman H. Phelps moved directly to Scranton.

Isaac Phinney, Gould's son, also moved to Scranton.

At Montrose, Isaac Post and his brother David Post established a business center and the Susquehanna county seat. A cousin, lawyer William Jessup, joined them.

Both communities, Montrose and Dundaff, were typical of Susquehanna County; they were dependent on farming and land development as primary sources of revenue.

A pamphlet advertising the joys of settling in America, specifically Susquehanna County, was read by Secku Meylert in London, England in 1819. Meylert was a political exile from the German Duchy of Westphalia. Meylert followed up on the promises of the pamphlet. In Susquehanna County, he found a wife, had children, and held the offices of chief clerk, deputy registrar and recorder of deeds.

Meylert also, from 1830 until his death in 1849, "had charge of nearly all the land estates belonging to non-resident land owners in northeastern Pennsylvania." Meylert was principally in "charge of the Meredith lands, and was succeeded by Michael Meylert," his son. The tracts of land that Samuel Meredith and brother-in-law George Clymer had purchased, starting after the War of 1776, in western Virginia, eastern Kentucky, Delaware and Sullivan Counties, New York, in addition to a goodly portion of northeastern Pennsylvania, totaling almost 2 million acres, were now managed by the Meylerts. (When the Meredith and Clymer lands were partitioned, Meylert acted for Meredith heirs only.)

The Meylerts, first father then son, became like Joseph Fellows, Jr. Just as Fellows had done, remaining a scrupulous agent for another's interest while at the same time making important personal investments from his own earnings, so Secku Meylert began purchasing land in his own name. Starting in 1846, Secku arranged to buy 76 tracts of land in the eastern portion of Lycoming County totaling approximately 33,000 acres. The following year Meylert's land lay at the center of Sullivan County (erected in 1847). After his father's death in 1849, Michael Meylert continued on, sometimes with the help of William Jessup.

This was the same period (1848) when Joseph Fellows dispatched his nephew Sylvanus Heermans to western Virginia, (now West Virginia) "bringing out a colony of Pennsylvanians and New Yorkers. He came as the agent of his uncle, Joseph Fellows, of New York, to look after his interest in a large body of lands in this [Preston] county.

"This town he called Fellowsville, in honor of his uncle. A land agency was established, and a hundred town lots laid out and offered for sale. Sites were selected for a large flouring mill, an extensive axe and scythe factory, an iron foundry, and a woolen factory."

Next, Joseph Fellows sent brother-in-law John Heermans to Kentucky to replicate the process.

By 1851, Horatio Pierce, with Thomas Gillespie, was buying Meredith lands in Wayne, Susquehanna, and Luzerne Counties from Michael Meylert. Pierce's direct land dealings with Meylert would continue until 1866.

And in 1853, Erastus Corning was on his way to northern Michigan.

In 1853, Joseph Fellows, Jr. was 71 years old. The men who depended on him, the men who surrounded him, the men who benefited from his actions were all wondering how much more he could do?

Chapter 42

The new economics of wealth and power, the Industrial Revolution, as viewed on the horizon of world economy, could only succeed with the help of the coal fields of northeastern Pennsylvania.

By the 1840s, very few men were living who had fought in the War of 1776. The number of American colonists who had been born to bow to their liege lords was also growing smaller. After all, the American Revolution had begun over 60 years ago.

By the same token, passage of time, the doctrine that wealth and power arose only by land ownership had encountered something really new on its horizon, an economic thought so different as to be revolutionary: land ownership would be just one of many paths to wealth and power. At the end of the first half of the 19th century, the full value of the steam engine was just beginning to be understood.

(One of the greatest repositories of the American settler's culture and civilization is the Winterthur museum and research center near Wilmington, Delaware. Its founder, Henry DuPont, when setting up the museum, generally limited its scope to the years 1607 to the 1840s. DuPont said that after the end of that period too many of this country's arts and crafts were machine made.)

By the 1840s, the people of northeastern Pennsylvania had demonstrated that the most efficient operation of a steam engine resulted from the use of anthracite coal. They had proved that a better quality of life in the home could be achieved by using anthracite. They had also shown the manufacturing capabilities of anthracite for producing iron and steel.

The new economics of wealth and power, the Industrial Revolution, as viewed on the horizon of world economy, could only succeed with the help of the coal fields of northeastern Pennsylvania.

And the people of northeastern Pennsylvania were at last ready to meet the demand - they knew where to find anthracite and they were equipped to get it to market.

By the end of the 1840s, it was becoming obvious that rail transportation would soon outdistance canal transportation; the New York Legislature had directed that a survey be made for a road from New York City to Lake Erie, just as the Erie Canal was nearing completion in 1825. Between then and 1840, several efforts were made to build a railroad across the southern tier counties of the state. The New York and Lake Erie Railroad Company became the New York and Erie Railroad (common usage called the Erie Railroad) and completed the first portion of its track, 46 miles from Piermont to Goshen, in 1841.

The Erie had been chartered in 1832, but work on the road itself did not begin until 1836, and lagged for a while until 1838 when an influx of money allowed construction to resume.

Until 1838, the D&H Canal Company and canal boat operators on the North Branch Canal at Wilkes-Barre had a monopoly on carrying anthracite from the northern coal field. News of the Erie's progress north and west worried the D&H particularly. The

D&H was assuming that Erie builders were heading straight for the Lackawanna River through Cobb's Gap (today's junction of Routes I-380, I-84, and I-81), and the coal that could be found there, on their way to Lake Erie.

In 1836, two companies were incorporated in Pennsylvania: The Pennsylvania Coal Company and The Washington Coal Company. Pennsylvania Coal could hold land in Pittston, north of Wilkes-Barre, and could construct a railroad from its land to connect with other transportation. Martin Curtis, Charles Fuller, James W. Johnson, Richard D. Lathrop, Charles T. Pierson, and Lewis S. Waters (Watres) were listed as directors. Washington Coal's land was limited to the Lackawanna River Valley with the same arrangement for its railroad, and its charter actually named the D&H as its connecting railroad. Its directors were Robert T. Conrad, William H. Dimmick, Martin W. Dingham, William Overfield, John R. Walker, Samuel G. Walker, and James Watson. Both companies remained paper companies.

As the Erie appeared to be recovering from another difficult financial period in 1845, a relative of the Wurts began buying land in Providence Township, lying in a northwestward line from Cobb's Gap (this was the same time period that the Scrantons were trying to get their iron works going at Slocum Hollow). When the Erie applied for permission to enter Pennsylvania in 1845, Legislators yielded to the request of one of their state Senators, William H. Dimmick, who was also legal counsel for D&H, and defeated Erie's bill. This put the Erie at a standstill.

The next year (1846), the Luzerne and Wayne County Railroad Company was formed. The directors of all three companies, the Luzerne & Wayne, Pennsylvania Coal, and Washington Coal, included names of people actively associated with D&H operations, and of course, a Wurts relative. Now, D&H plans were finalized. When the Erie again applied to come into Pennsylvania, it could have its approval.

As for the D&H, yes, it did want to keep the Erie out. (After all, what business person wants to share profits when he's been accustomed to keeping them all for himself?) But the D&H was also considering the future when it involved itself in organizing these three companies. The future for the D&H centered around a charter that was supposed to expire in 1852, and incontrovertible evidence that demand for anthracite would increase at a spectacular rate each year. From 1842 to 1850, D&H made extensive improvements to both its gravity railroad and the canal to increase capacity. One of the improvements brought John A. Roebling in 1846 to the banks of the Delaware River, where he studied the terrain at Lackawaxen prior to designing his style of suspension bridge for an aqueduct over the river. (Roebling's bridge is still standing, and in use today, and is listed in the National Historic Register. Roebling is best known for designing the Brooklyn Bridge.)

However, improvements to the canal and railroad were not equalized. The canal was enlarged to carry larger boats for a yearly total of at least one million tons of coal. The railroad was brought up to where it could carry a half-million tons of coal a year. Clearly, additional anthracite, or some other commodity, would have to come from somewhere to fully utilize the larger canal. D&H president John Wurts reminded the Board of Directors of this fact at their December 31, 1846 meeting.

The D&H faced another threat, too. 1847 was the year that George Scranton offered stock in a proposed "locomotive engine road" from Great Bend (in Susquehanna County) to the Delaware Water Gap.

The three companies, Pennsylvania Coal, Washington Coal, and Luzerne & Wayne County R.R., would make for a powerful grip on all coal and transportation out of the area north of Wilkes-Barre, from where the Lackawanna River emptied into the Susquehanna River (at Pittston) all the way up the Lackawanna to Carbondale. The D&H was ready. As for the Scrantons selling stock, D&H knew all about paper companies; the threat of competition was there, but it wasn't yet as real compared to the Erie.

But suddenly, something was wrong - the Erie was heading for Port Jervis! (It reached there on December 31, 1847.) Now it was "another ball game." The D&H realized that the Erie was not going to challenge it to a head-on confrontation. D&H put its plans, and extra companies, on a "back burner." At its July 23, 1846 Board of Directors meeting, a million dollar liability just to get Washington Coal Company going was considered - and all action was tabled.

D&H's contingency plan was too good to be allowed to go to waste. Some D&H stockholders, on their own, decided to "breathe life" into the various companies. William Wurts, with others, purchased Washington Coal Company's charter for themselves. D&H chief engineer James Archbald (who had worked on the Erie Canal) was installed as president.

The D&H not only commented on, but seemed to applaud the action surrounding Washington Coal Company. The Annual Report issued on March 28, 1848, read in part: "The most important event, however, that has occurred during the year, in reference to the interest of the stockholders, is the commencement of a railroad by the Washington Coal Co., which is intended to connect the lower part of the Lackawanna coal field with the canal of this company about ten miles below Honesdale. It is contemplated by those engaged in making this road to have it finished and in use by August 1849."

With its once controlled paper companies starting to come into their own, the D&H again moved to cover all bases - on April 28, 1847, the Managers established a set of tariff rates on coal "to create motive and inducement for capitalists to invest their funds in the construction of a railroad or railroads, to be connected with the canal of this company at or near the mouth of the Wallenpaupack River."

(This could have been meant for the Scrantons and their shipments of iron, too.)

But by June 24, 1847, the picture was again shifting. Who knew what companies were going to ship where? Another mix of D&H

stockholders organized The Wyoming Coal Association, in New York City. Like Washington Coal and Pennsylvania Coal, it could have a connecting railroad - to connect to the D&H.

Meanwhile, Pennsylvania Coal Company, left to its own devices since its inception in 1836, had managed under its president Charles T. Pierson to open one mine shaft in 1842 at the corner of Railroad and Main Streets, in the center of Pittston. (As you can see, not too much capital was going into this project.)

As the year 1847 progressed, and the Erie R.R. moved closer and closer to Port Jervis, ignoring Lackawanna Valley, the D&H became more and more disenchanted with all its plans. Archbald took the lead, and this time, acting as chief engineer for Pennsylvania Coal, laid out a gravity railroad track from Pittston to the juncture of the Lackawaxen River, Middle Creek, and D&H Canal. There, a flat level space was selected for a coal basin. *Today*, we know the coal basin as Bingham Park and the area as the Borough of Hawley. (New Yorker Irad Hawley was one of the D&H directors. He also served as Pennsylvania Coal Company president for one year, 1850-1851.)

The railroad had a slow start. The first shovel broke ground in 1847, but it was not until May of 1850 that the line was completed. Along the way, starting April 1, 1849, all the other companies instituted by D&H were transferred/sold to Pennsylvania Coal, including Wyoming Coal on May 30, 1851.

After that, Pennsylvania Coal showed it was thinking about becoming a major player in its own right when it came to transporting anthracite coal out of the mountains of northeastern Pennsylvania. By 1853, Abram S. Hewitt, president of Sussex Railroad in New Jersey, was authorizing his engineers to assist in surveying a road across New Jersey to New York City for the Pennsylvania Coal Company. (Pennsylvania Coal Co. was in the process of acquiring a New Jersey paper railroad, the New Jersey, Hudson & Delaware. The project never got off the ground.)

Dr. Benjamin Throop likened Pennsylvania Coal to "a kind of mutual company."

But just how much was Pennsylvania Coal a mutual company?

Archbald's assistants in laying out the path of the railroad track were William Maffat and Gideon Frothingham. Maffat was responsible for that portion of the line from the Moosic Mountain tunnel west to Pittston. Frothingham surveyed the road from the tunnel east to its terminus at Hawley, including a portion along the southern edge of 9,271 acres purchased by Joseph Fellows, Jr. almost 20 years earlier.

And Pierson, after he resigned as president of Pennsylvania Coal in 1848, teamed up with Horatio S. Pierce to buy 176 acres of coal land in Pittston Township.

Archbald's assistant engineer James Clarkson, along with two other D&H personnel, John Hosie and Alexander Bryden, who had all moved over to Pennsylvania Coal, joined not only Horatio Pierce but Thomas Gillespie in 1852 to buy more coal land: 88 acres in Pittston. Again, two years later, in 1854, Pierce and Gillespie, with Pennsylvania Coal's Clarkson, Hosie and Bryden, bought nine tracts of land in lower Luzerne County from Alexander Gray.

This last transaction, involving approximately 4,000 acres, was a large one, and Gray remained as a one-sixth owner. Twelve years later in 1866, still more D&H men teamed up with Pierce, Gillespie and the others for even more land. (Remember, all the land they were buying was underlaid with coal.) Thomas Dickson came in as an additional partner because he had land to add. Other new partners were John S. Law, William Law, John Howell and Andrew Bryden on Alexander Gray's death. Dickson would become president of the D&H in 1869.

Chapter 43

Now that land was no longer the only sign of wealth and social standing, it was treated as any other commodity - it had value if it had use.

While Horatio S. Pierce was in Pittston buying land, William W. Winton was in Providence Township both buying and selling land. Winton had gotten his entré into land dealings first through his father and then in administering his father-in-law Henry Heermans' estate.

Winton's father, Andrew Jackson Winton, had been active in buying and selling land throughout the first quarter of the 19th Century, in Otsego County, New York. He and his brothers took part in over 60 land transactions while living there. For the most part, sales involved relatively small parcels. Otsego County was, and still is, excellent farm land. (While Winton aggressively dealt in land, Horatio Pierce's father, Levi Pierce, was more modest, content with his one home tract.)

As for Henry Heermans, after moving his family to Providence in 1828, he continued to acquire more land in the area, always with an eye for coal. Heermans named William Winton executor of his estate in his will. On Heermans' death in 1843, Winton turned division of the real estate over to court-approved appraisers. But when the heirs took exception and asked to be bought out, and Heerman's widow, with dower rights, remarried only to divorce shortly afterwards, Winton not only had to administer the estate but also became an owner in his own right of some of the estate lands.

Winton purchased more land in Providence Township, most of it adjoining Heermans' property. At the same time, he took Charles J. Atwater into partnership in the general merchandise store that he had bought from his father-in-law and moved to the intersection of North Main Avenue and West Market Streets. (Winton's relationship to Atwater took on added significance when in 1854 Atwater's daughter Ellen married Charles Law. The Laws were active in Carbondale and the D&H Canal Co., and Charles' brother John S. Law would join Horatio S. Pierce in acquiring coal land for Pennsylvania Coal Co.)

While Winton was becoming a large landholder in the Providence Borough section of Providence Township in the late 1840s, the Scrantons and their iron making venture were just beginning to fly.

William E. Dodge, a merchant in New York City, and business partner and son-in-law to Anson Phelps, of the firm Phelps Dodge & Company, had enough clout by 1847 to be on the board of directors of the New York and Erie Railroad. It was Dodge, along with Erie President Benjamin Loder, who visited the Scrantons at their mill in Slocum Hollow to consider the Scrantons offer to manufacture rails for the Erie.

"We were greatly pleased with what we saw," Dodge later recalled, "and particularly with the prospective value of their enterprises, but concluded they could not arrange their works in time for us [Erie R.R.], and fearing a disappointment the proposition was declined."

George and Selden Scranton's persistence carried the day, however, and, after meeting with Dodge and others in his New York office, were advanced enough money to start making the "T" rails for the Erie. Once the Scrantons began delivering the rails, confidence soared, and there was general recognition among the men directing the operation that the Erie would get to Binghamton, New York on time (by January 1, 1849). The Scrantons and the men who had backed them to this point looked ahead to other possibilities.

But more money would be needed to make the whole dream come true - as William Henry had hoped - a self-sufficient community with its own links to the rest of the world.

The stock book was opened in 1847 for subscriptions for a "locomotive engine road from Great Bend to the Delaware Water Gap." The purpose was to connect with the Erie. By November 1, 1848, it was apparent something more had to be done to complete the dream of connecting a wild, isolated forest land with civilization.

All of the partners of Scrantons & Platt, both general and special, entered as limited partners, into an agreement whereby Selden Scranton, who had been acting as deeded title holder to almost 7,000 acres of land in northeastern Pennsylvania, would transfer the properties to William E. Dodge, John Howland and George W. Scranton as a trust. By the agreement, the trustees would be given the power to use the trust assets as backing for any corporation formed within the following ten years that involved the limited partners. Dissolution of the trust would return assets, in proportion to their investment, to the individual partners.

Four months later, on March 19, 1849, after the Scrantons purchased the Liggett's Gap Railroad charter, the Commonwealth again gave approval for corporate activities. Thomas Meredith, one of his father Samuel Meredith's heirs to all the Meredith lands, had been the original force behind Liggett's Gap in the

1820s and 1830s. At one point, in 1832, Henry W. Drinker was president and William Henry and nephew John Jordan were treasurer and secretary. But the sale of stock in New York City had not been successful and activity on the railroad languished. When the Scranton group acquired Liggett's Gap, a survey and estimate was ordered. To build a railroad from the iron works at Scranton (then called Harrison) north through Liggett's Gap to a junction with the Erie at Great Bend in Susquehanna County, a distance of approximately 53 miles, an investment of $1.6 million would be required.

In April 1849, just a few weeks after purchasing the Liggett's Gap charter, those interested in the Scranton iron works generated two corporate actions. (Additionally in 1849, George W. Scranton and William E. Dodge purchased the Cayuga & Susquehanna Railroad as a way of linking Liggett's Gap, via Erie and Cayuga, to the Erie Canal.) One corporate action was to amend the charter of Liggett's Gap R.R. to allow for ownership of up to 1,000 acres of coal land. (The first president of Liggett's Gap was not elected until January 2, 1850. His name was John J. Phelps. Yes, that's right, the same John Phelps who had married into the Phinney family, the developers of Dundaff in Susquehanna County.)

As the building of the road progressed northward from the iron works, behind Main Avenue, and across Liggett's Creek, the charter was amended again - this time (April 14, 1851) for a name change. Liggett's Gap R.R. became the Lackawanna & Western Railroad Company. The first train to run between the Erie connection at Great Bend and Scranton met the Erie on October 16, 1851. The Erie train was out of New York City and was carrying all those important personnages that make for a grand opening celebration. Three hours later a large crowd of well-wishers welcomed the New Yorkers to Scranton.

The second corporate activity in 1849 (on April 7, 1849) was to obtain a charter for a new railroad company: the Delaware & Cobb's Gap Railroad. Delaware & Cobb's Gap could not only lay track from the Stroudsburg area to the iron works at Scranton but also spur trackage to meet the Erie R.R. near Port Jervis, New York, as well as to the Belvidere-Delaware R.R., in New Jersey, and also build as many bridges as it wanted across the Delaware River between the Delaware Water Gap and Blairstown, New Jersey to the south. However, where Liggett's Gap had sold enough stock shares this time, and had been able to establish corporate officers in January 1850, it was not until December 4 of that year that the Delaware & Cobb's Gap was at the same stage of development. On December 26, at a meeting held at Stroudsburg (Monroe County), George W. Scranton was elected president. And yes, John J. Phelps was a director, too, as were William E. Dodge, John I. Blair, Seldon T. and Joseph H. Scranton, Joseph C. Platt, and Philip H. Mattes. Added to them were Horatio Nicholson, a Heermans/Winton first cousin, and Stroudsburg residents James M. Stroud and Stroud Holinshead among others.

The hard financial facts of building a railroad north to Great Bend could not be ignored. An estimate of $1.6 million had turned into a reality of $2.1 million, plus an additional half-million dollars for rolling stock, tools, machinery, etc.

So, more money would have to be brought in for the track southeast from Scranton. It was done in much the same manner. This time, Selden T. Scranton, along with John J. Phelps and John I. Blair were the ones to hold over 31,000 acres in Luzerne, Monroe and Wayne Counties in a five year trust. The beneficiaries were William E. Dodge, Moses Taylor and Samuel Marsh, of New York City, and John I. Blair, of Warren County, New Jersey.

The way west, however, remained like an army's exposed rear flank. There were no railroads marked from Scranton to carry anthracite into the interior of Pennsylvania. Less than six months after the Liggett's Gap had connected with the Erie at Great Bend, a group of 28 men incorporated the Lackawanna and Bloomsburg Railroad Company. This new railroad would originate at Scranton and end at Bloomsburg, in Columbia County. It also had the chartered right to extend west to Danville, in Montour County.

The Bloomsburg Railroad was not started by the Scrantons, Dodge, Phelps, Blairs, etc. But of the 28 men involved, one was Horatio Nicholson, director/shareholder in Delaware & Cobb's Gap, and another was Dr. Andrew Bedford, an incorporator of Liggett's Gap.

The men who would soon form the Delaware, Lackawanna and Western Railroad (DL&W) were described in 1852 and afterwards as "interested" in the activities of the Bloomsburg line, so much so that in 1856 four Lackawanna Iron and Coal Company men, Selden Scranton, Samuel Hoyt, John Brisbin, and James Archbald, were named to the Bloomsburg's board of directors. But it was not until 1872 and 1873 that the DL&W had enough capital to acquire the Lackawanna and Bloomsburg line outright.

From the time it was chartered, it took the Lackawanna & Bloomsburg until 1857 to reach Bloomsburg, and until 1859 to reach Danville. While it "covered the backdoor" for the DL&W, the Bloomsburg line served an even more important purpose - "almost every foot of this road runs over coal, or iron, or limestone..." Not only had the DL&W had its interests protected when the Lackawanna & Bloomsburg was built, but the future Lackawanna Iron and Coal Company (LI&C) was guaranteed a supply of top grade iron ore and limestone.

The building of the road to Bloomsburg more or less parallelled that from Scranton to Delaware Water Gap. But the Delaware & Cobb's Gap and the Liggett's Gap roads would constitute the original DL&W R.R.; the Bloomsburg would always be classified as a "division of the DL&W."

Before actual construction began on the Delaware & Cobb's Gap, both it and the Liggett's Gap had their charters amended, combining the two railroads into The Delaware, Lackawanna and Western Railroad Company (DL&W) as of March 10, 1853. Now, construction could begin on the way southeast.

One more corporate act was yet to be completed - incorporation of the partnership firm of Scrantons & Platt, the great iron works. Three months to the day after the birth of the DL&W R.R., the

LI&C Co. (Lackawanna Iron and Coal Company) was born on June 10, 1853.

Gone forever were the days of the private individual as the big landholder. No more Tripps, Slocums and Griffins owning almost three-quarters of Providence Township. The seemingly infallible tenet that land was the only indicator of a person's wealth, power and social position had at last been shaken loose from man's psyche.

The age of John Nicholson and Robert Morris was gone. In a way, they had foreshadowed the future when they formed their North American Land Company in 1795 to mitigate debt demands. The Pulteney Associates was also of another age. But having passed into estate status, it managed to survive. Perhaps estate administration was close enough to impersonal corporate structure so that trustees could slip silently from one time to the next. Stephen Girard, Samuel Meredith, George Clymer, and William Bingham all continued on, more active as estates than when the originators were alive. Bingham lands in Tioga County were actively developed as soon as the Corning Company men had opened up the route for Blossburg coal to flow northward.

For the most part, new land owners would be impersonal corporate entities, machine-like in the way of the creations of the Industrial Revolution, corporate officials meshing like any well-oiled set of gears. In the beginning, individuals would constitute a layer hidden underneath the corporate structure. These individuals could claim land ownership; they had invested heavily in the corporation. But as time went on into the 20th Century, such players would almost all disappear. In their place would come not other like-minded individuals but banks and financial houses with their mortgage and loan contracts.

The anthracite transportation companies led the way for the corporations' domination as large landholders.

From the very first days of the Liggett's Gap R.R., George Scranton had spoken of the DL&W as a locomotive engine (as opposed to gravity) road. Locomotives burned a variety of fuels for energy. The first was wood, but wood, as in iron furnaces, required vast timber tracts for supply. Bituminous soft coal became an alternate fuel, especially for railroads in England. When it came to "getting up a good head of steam," there was nothing, however, that produced the energy generated from burning anthracite. By 1850, anthracite was an acknowledged superior alternate for driving those big locomotive wheels - so much so that in 1853, the DL&W "set about constructing a model engine of 'great power' for burning anthracite coal."

Official corporate policy stated, "should the anticipations in this respect [the prototype anthracite-burning steam engine] be realized, we design at once to order several more of the same character. Coal-burning engines are now in successful operation on the Reading [Philadelphia and Reading R.R. tapping Schuylkill County coal fields] and other roads, and it is believed that they will soon come into very general use, thereby causing a large increase in the consumption of anthracite coal, and at the same time effecting a vast saving in the cost of fuel, no small item in the running expenses of all railroads."

First the stationary engine, now the portable engine had been adapted to handle the tremendous steam pressure created by the hot heat of anthracite. The steam engine and anthracite were together at last. The United States was set on a path that had only one outcome - to be a world power.

Chapter 44

Everyone knew there was coal there, somewhere in the Mid-Valley. After all, the D&H had found enough to open a mine... But where...?

There still remained one section of the Lackawanna River valley to be accessed: Carbondale to Honesdale was covered by the D&H and its gravity railroad; Pennsylvania Coal Company could transport from Scranton to Pittston; and the DL&W from the center of original Providence Township outward. All three companies linked to other transportation to reach world markets.

But the space between Scranton and Carbondale, where anthracite was known to exist, was still untouched.

The D&H was the first to come into the Mid-Valley. It built gravity tracks as far as Archbald (about six miles south of Carbondale) where it owned some coal land. But the D&H had a problem; its original charter had been for 30 years. It was just possible that the Commonwealth of Pennsylvania could take it over, under its charter. Until 1852, the D&H would walk a narrow tightrope.

In fact, it improved its assets but it did not stray from its chartered rights. Starting in 1842, D&H Superintendent James Archbald authorized major changes on the gravity track, enlarging the canal to the Hudson River, and contracting with John Roebling to build three suspension bridges as aqueducts, across the Delaware and Lackawaxen Rivers. The D&H did not show steam engines as operating anywhere on their system.

All that changed after the Pennsylvania Legislature determined that $1.2 million plus would have to be handed over to D&H stockholders if Pennsylvania were to assume ownership control. Taxpayer dollars were not to go for this. On April 30, 1852, "the privileges and franchises of the company were made unequivocally perpetual."

With the question of its charter existence resolved, the D&H began adding steam locomotives to its system. The road going down the Lackawanna Valley slowly increased in mileage, first to Olyphant in 1858 and to Providence Borough/Scranton in 1860. The newly built trackage could handle locomotive engines.

To help the D&H expand, a former employee (and the son of a former employee), Thomas Dickson, set up Dickson & Company in February, 1856 at Scranton. Two brothers, John and George Dickson, plus members of the Wurts family, joined Thomas Dickson. Also joining Dickson were Charles T. Pierson (Horatio Pierce's co-land owner in Pittston) and Joseph Benjamin, once employer and co-partner of Dickson in a foundry and machine shop at Carbondale. The new company's first contract in 1856 was for the building of engines and boilers for D&H's new railroad over Moosic Mountain. In 1862, the company changed its name to The Dickson Manufacturing Company and purchased Messrs. Cooke & Company (of Cliff Street, Scranton), manufacturers of locomotive steam engines. The first four railroad steam engines used by the D&H on its tracks between Scranton and Carbondale were all made on Cliff Street. The first three were produced as Cooke engines, and the fourth bore the Dickson name. (Dickson would build 197 locomotives for the DL&W by the year 1899.)

As for the Mid-Valley, other than the D&H's track to its coal mine in Archbald, the Mid-Valley remained quiet and undeveloped, recognized solely as a source of timber.

Everyone knew there was coal there, somewhere in the Mid-Valley. After all, the D&H had found enough to open a mine... But where...?

One person who knew where coal could be found was the engineer James Seymour. He located coal in 1854, near the Lackawanna River just south of Archbald. Seymour found partners for his plan to develop a coal mine. They were Samuel Meredith heirs, represented by Thomas Meredith and Meredith agent Michael Meylert; they were also the Honorable William Jessup, at one time a judge for Luzerne, Susquehanna and Wyoming Counties. The area would be called Jessup.

Some Meredith heirs, especially Thomas Meredith, had been trying to open a road or railroad through the Lackawanna valley as early as 1826. It was chartered as the Lackawanna and Susquehanna Railroad, and Seymour surveyed for a road from Liggett's Gap to Great Bend, Susquehanna County, in 1830. But Meredith never seemed to be able to generate enough interest in stock sales. For years, the Lackawanna & Susquehanna remained a "paper railroad." When Seymour was finally able to pinpoint a coal location in 1854, the picture changed dramatically.

Referred to as the Lackawanna Railroad, "sometimes called Lackawanna and Susquehanna railroad," the charter was extended on April 20, 1854 for another ten years starting in 1856. Thomas Meredith put all his ownership interest in Meredith coal lands into the Lackawanna Railroad through Michael Meylert. Then, on September 19, 1854, William Jessup, as president of Lackawanna Railroad, entered into an agreement with Michael Meylert, Christopher Ward, and Henry Doolittle, as general contractors, for the building of a roadbed and laying of tracks from Scranton (Nay Aug/Greenville section) to Meredith/Childs in Carbondale Township. The price tag was $28,000/mile.

So far, the Lackawanna Railroad was the creation of a group of men who had their roots in Susquehanna County. Two other men, also with ties to Susquehanna County, played a less publicized role. David Wilmot, born in Wayne County, elected to U.S. Congress to represent Bradford, Susquehanna and Tioga Counties, authored the Wilmot Proviso in 1845. The Proviso did not allow slavery to exist in any territory acquired from Mexico. Farris B. Streeter, judge and lawyer like Jessup and Wilmot, served for a while, starting in 1853, as solicitor for the U.S. Treasury.

What does all this have to do with the Mid-Valley and Jessup?

Wait....

Antonio Santa Anna, victor of the battle at the Alamo, Texas, in 1835-1836, led Mexican forces against the United States in the Mexican War of 1846. At the end of the War, U.S. President James Polk requested an appropriation of $2 million to indemnify the Mexican government for its loss of Texas. Wilmot's Proviso was an amendment to the appropriations bill. When the bill was finally passed in 1847, the Proviso had been deleted. Santa Anna, for his part, once more assumed the presidency/dictatorship of Mexico from 1853 to 1855, when he was again overthrown.

This last time around, Santa Anna was president of Mexico just long enough to buy a very convincing sales pitch - after all, a lot of money had come into the Mexican treasury not too long ago - that Jessup, Luzerne County, Pennsylvania was just the place for an investment.

Yes, that's right. You've read correctly. That section of Moosic Mountain in the Mid-Valley was on its way to becoming a Mexican colony.

Published reports placed Santa Anna's investment in the Lackawanna Railroad project at $125,000. But from land prices, it was probably much more than that. In the first three months of 1855, the estate of Maria Meredith, in addition to William Jessup, John Torrey, Benjamin Reed, Wellington Tyler and Benjamin Huntting sold land to the Lackawanna Railroad along the planned path of the road for a total of $356,610.50. The cost of constructing the road was $28,000/mile and the nine miles that were actually built, from Jessup to connect with DL&W at Nay Aug/Scranton added up to an investment of a quarter of a million dollars. Even more money went into the project. Two mine shafts were opened, two engine roundhouses (roundtables) were constructed, one at Jessup and one at Nay Aug, also all the machinery and necessary housing for opening a mine, hoisting and preparing coal for market, including a breaker, water tower, offices, weigh scale foundry, blacksmith shop, railroad engine and cars, and at least seven miners' houses.

An agreement was reached with the DL&W that when it reached tidewater at Elizabeth, New Jersey, the Lackawanna Railroad would have its own storage facilities. And then there was Jessup's Wall Street office in New York City. Minor expenses weren't so minor when totaled together. Timber from adjoining land owned by Hollenback heirs and from Lewis S. Watres, who lived between Jessup and Archbald, soon reached $1,000; there were incidental expenses to Theodore F. Hunt, Dr. Benjamin Throop, and others, too.

All this added up to a pretty impressive sum for the year 1855.

Would the Mexican community have ever become a reality? Who knows? Santa Anna fled to safety in the Caribbean Islands and returned to Mexico in 1874, penniless.

And the Lackawanna Railroad?

Well - an unpaid debt caused the whole system, including the coal land, to be sold at sheriff's sale on February 29,1860. The debt was to Horatio S. Pierce and Thomas Gillespie. The purchasers were Andrew McClintock and George Byron Nicholson. (McClintock had married a Matthias Hollenback heir, and Nicholson, like his half-brother Horatio Nicholson, was first cousin to Catherine Heermans Winton and her husband William W. Winton.)

But the story does not end here.

A year later, Nicholson sold his interest to a newly incorporated company, the Grassy Island Coal and RailRoad Company, with Horatio S. Pierce as president. The railroad was not kept up.

Chapter 45

For northeastern Pennsylvania, the 1850s were perhaps the most crucial years, when everything came together that would have a bearing on the future.

The first coal land the Delaware, Lackawanna and Western Railroad Company (DL&W) developed was located along its tracks in what is known today as West Scranton, in an area just south of Isaac Tripp's house, standing from the late 1700s. The mine was dubbed the Diamond Mine. The first shovel was turned in 1851 by John Jermyn, mining engineer and contractor. Jermyn supervised workings at the Diamond Shaft for about five years. 1851 was also the year the Erie Railroad reached Buffalo, New York and Lake Erie; coal cars leaving Scranton had direct access to communities along the Great Lakes and to the west.

When Jermyn left the DL&W and the Diamond in 1854, it was to go to work for the New York and Pennsylvania Coal Company. In Providence Borough, now North Scranton, Jermyn opened the Rockwell mine on the banks of Liggett's Creek.

And just who was the New York and Pennsylvania Coal Company?

While Horatio S. Pierce was busy at Pittston with Pennsylvania Coal Company and William Jessup with the Lackawanna Railroad, William W. Winton was active in Providence Borough. He, his father Andrew J. Winton, his brother Andrew Winton, and two others, Charles Rockwell and Daniel Botsford, incorporated the New York and Pennsylvania Coal Company.

"In this undertaking he [Jermyn] was engaged some four or five years and was highly successful. The contract having been filled and the work accepted by the company in the year 1859," Jermyn left to work elsewhere.

After the Wintons closed out the mining operation, Charles and Henry Rockwell took over, leasing it to the Delaware & Hudson Canal Company. This gave the D&H a foothold in Scranton and apparently established the pattern of the two companies (D&H and DL&W) mining coal deposits side-by-side without conflict.

William Winton had also acquired a brother-in-law: Abram Beemer Dunning. Dunning had worked with Winton in what was once their father-in-law's store in Providence. In 1848, Dunning married Catherine Heermans Winton's sister, Mahala Heermans. It wasn't long thereafter that the family found what Dunning was best suited for - politics. "Thrice elected - a compliment seldom paid in this county [Luzerne, now Lackawanna]," Dunning was sent to Harrisburg as a member of the House of Representatives in 1852, 1853, and 1854. These were the crucial years when the DL&W was expanding, when Scrantons & Platt became LI&C, when the D&H charter was made perpetual, when the Lackawanna Railroad was opening up Jessup, and when the Lackawanna and Bloomsburg R.R. got its right-of-way as far as Danville in Lycoming County.

As Dunning's last term in office neared its end, Michael Meylert representing another landowner (Cremond/Walker), managed to find 414 acres, partly in Madison Township and partly in Jefferson Township (Lackawanna County near the Wayne County line) to sell to the Dunning family. Eventually, the DL&W would include the village of Dunning (now Elmdale) as a stop on its way to the Delaware Water Gap; and a quarter million dollar tannery would operate nearby on the land leased from Dunning.

Once out of office, Dunning took over Winton's land interests in Pittston and Lackawanna Townships (now Moosic) in 1857. The tracts were rich in coal and a portion had at one time belonged to Henry Heermans. Winton, however, had not owned them outright until he purchased the land from cousin Horatio Nicholson in 1855.

For northeastern Pennsylvania, the 1850s were perhaps the most crucial years, when everything came together that would have a bearing on the future.

One more thing remained to be accomplished in the 1850s. On May 10, 1855, the exchange and banking office of Mason, Meylert & Company opened. This was the Scranton area's first bank. The Meylert portion of the banking firm was Michael Meylert and his brother Amos (A.N.) Meylert. The Mason part was Gordon F. Mason, of Towanda in Bradford County.

In the 1830s and 1840s, the D&H had been the only source of cash and ready money east of Wilkes-Barre. Carbondale and Honesdale became cities dominating northeastern Pennsylvania because of the hard currency coming into the D&H. But once the Scrantons tapped into Erie Railroad funds, the situation changed. The D&H no longer had a monopoly on anthracite coal going to New York and New England; was no longer the single largest employer; but after it had perpetuated its charter, the D&H seemed to adjust quite well to "local competition." It was in the 1850s that major money supplies began to stockpile for the first time not only at Wilkes-Barre and Carbondale/Honesdale but also at Scranton, waiting for those who dared to claim some or all of it. And it was from the 1850s that the men came who would claim it - and continuing claiming it far into the future. Money never goes out of fashion; it is always desirable. Would these men, who boldly stepped forward laying claim to the riches flowing in to the world center of a source of energy, always be close to the height of fashion? Would they always be powerful while they lived? How could their power survive them?

No other interests or personalities ever succeeded in penetrating, in the sense of grabbing hold of, the flow of anthracite that fueled the Industrial Revolution. The power of anthracite was not that it was the only fuel, but that it was the best suited for turning the wheels of industry. It created an overabundance of energy that no other nation in the world could match.

Many of these men who successfully claimed the riches from the flow of anthracite all over the world linked moral values and religious fervor with the economic ideal. To them, it was an indication that man was becoming an ever better human being. In an age of ever better technological improvements, it was inevitable that man would apply the same logic to himself - to aim for an ever better, improved version of himself. Unfortunately, starting in the 1870s, some members of the religious community would take hold of this thought and invert it to an ultimate conclusion - that those who had been able to amass great wealth had succeeded only because they were leading a more godly life; that these people were more devoted to God, and personal wealth was a sign of approval from God. Collaterally, the thought followed that those who were poor and/or afflicted were living proof of God's disfavor, and an outward sign that they were not devoted enough to God.

In the first days of success, however, men like William Jessup and those with Quaker backgrounds adopted a very practical approach to their self-improvement program. They supported abolition. Quakers in the Oblong, in Dutchess County, New York, had gone on record as early as 1767 in favor of freeing slaves. The active abolitionists of Susquehanna County went on record in 1836, when 80 of them formed the Susquehanna County Anti-Slavery & Free Discussion Society. Article Two of the Society's Constitution, declaring the purpose of the organization, stated in part, "the total abolishment of slavery in our country by all constitutional and lawful means..." (David Wilmot had a lot of home-grown support for his Proviso nine years later.)

There were many stops in northeastern Pennsylvania and Pulteney's Genesee country (in New York) for those traveling the underground railway before the Civil War. The two rivers, Susquehanna and Delaware, were easy guides north to Canada and the western United States. There are many incidents on record involving irate slave owners searching Wilkes-Barre for their missing property. As for the abolitionists in Wilkes-Barre, hasty wagon trips to the next station at Providence Corners, or to an emergency temporary hideout in a mine tunnel, guaranteed continued safe passage to freedom. From the Delaware River, the "safe house" in Shohola Township (Pike County) with its secret room is still standing.

After Wilmot's Proviso failed in the U.S. Congress in 1846, its supporters coalesced into the Free-Soil political party at a convention in Buffalo, New York, in 1848. By 1852, Free-Soilers had increased their representation in Congress and because of the passage of the Kansas-Nebraska Act in 1854, had enough momentum to become the Republican political party. The first Republican Party national convention was at Pittsburgh, Pennsylvania, on February 22, 1856, and John Fremont, who had found gold in California, was put forth as the Republican candidate for U.S. president.

(The Kansas-Nebraska Act created Kansas and Nebraska. To attract southern votes for the bill, it also repealed that portion of the Missouri Compromise that had prohibited slavery north of Latitude 36°30'N.)

As the time grew closer to another presidential election, Republican party members began planning, as early as 1859. In May of that year, at Honesdale, in Attorney Samuel Dimmock's office on Ninth Street between Main and Church, four men met: Horace Greeley, Simon Cameron, Andrew Curtin, and Dimmock. They intended to nominate Abraham Lincoln as the Presidential candidate at the Republican national convention coming up in

1860 at Chicago, Illinois. When the Republicans convened on May 16, 1860, David Wilmot was named temporary chairman of the convention.

Jessup was a delegate to the convention. Afterwards, he told his family that he had spent all of the night before the voting in prayer. Jessup prayed for guidance as to whether or not to support Lincoln. In the morning he announced he would cast his delegate vote for Lincoln. Lincoln won and went on to become U.S. president in 1861.

The meeting at Honesdale the year before had been crucial to the success of the Lincoln nomination. Cameron was a United States Senator from Pennsylvania. As such, he would receive the presidential nod from Pennsylvania at the convention. Curtin's interest was as the Republican candidate for governor of Pennsylvania in the same election. His political future depended on who won the presidency. Greeley, for his part, had no political aspirations. But as editor of The New York Tribune, he could wield tremendous influence, especially as he had already begun writing editorials in favor of a Missouri Whig. (Greeley had become a familiar name in Pike County, starting in the early 1840s. He gave money and public recognition to an experimental community-living colony, Sylvania Association, near the junction of what are now Pennsylvania Routes 590 and 434 in Lackawaxen Township. Its members disbanded in 1843.) As for Dimmock, his political aspirations were rewarded in 1873 when he was appointed Pennsylvania Attorney General.

The 1850s were a time of expansion and development everywhere throughout the United States. One of the busiest was Upper Michigan. With statehood in 1836, Michigan had received the northern tip of Wisconsin Territory where it curled over the Michigan Peninsula. This section is still a part of Michigan, and today is commonly referred to as the Upper Peninsula. The State of Michigan had accepted the northern lands in exchange for relinquishing its claim to the Toledo Strip, an approximately seven mile wide strip of land at the base of the Lower Peninsula, which was given to Ohio and Indiana.

It was not until 1840 that Surveyor William Burt drove a stake into the ground at Sault Ste. Marie, establishing the northern point of Michigan's meridian from which all land in the state is measured. Then, in 1841, Douglass Houghton, who had been exploring the Upper Peninsula for ten years, acknowledged the presence of large deposits of almost pure copper on the Keweenan Peninsula. (The Ontonagon Boulder, containing two tons of pure copper, is in the Smithsonian Museum in Washington, D.C.) The United States and Michigan finally recognized that perhaps untold amounts of mineral riches lay hidden on the Upper Peninsula. So, in 1844, Houghton and Burt were tasked with making a joint federal and state survey of 4,000 square miles of the Upper Peninsula "with reference to mines and minerals."

It was on the morning of September 19, 1844, that the survey team noticed erratic movements of the compass needle. "As we looked at the instrument, to our astonishment the north end of the needle was traversing a few degrees to the south of west," Houghton's brother Jacob recalled later. Team members soon found an abundance of iron ore samples nearby. Named the Marquette Range, this was the first of the great iron ore deposits discovered across northern Michigan, Wisconsin, and Minnesota (the Mesabi Iron Range is at the western end of the ore deposits in Minnesota).

Demands for Marquette Range ore and copper skyrocketed to the extent that in one year (1855) alone, 1,449 tons of iron ore and six million pounds of copper were readied for market. As the commercial need for iron and copper had increased each year, transportation had just as quickly become a major problem. Water transport is the most economical of all modes for moving heavy bulk around. The Upper Peninsula was ideally suited for water transport. Cities such as Chicago, Milwaukee, Detroit, Erie and Buffalo were as reachable as next door neighbors, via the Great Lakes. But there was one bottleneck in the water route - between

Lake Superior and Lake Huron. A series of rapids and falls in the St. Mary's River channel, which separates the United States from Canada near the eastern tip of the Upper Peninsula, forced all travelers onto land. This was known as the Sault Ste. Marie portage.

The Northwest Fur Company, as early as 1798, had tried to overcome the portage problem by building a canal with a lock on the Canadian side. This lock was destroyed in the American-British War of 1812. It was not until the need arose to take out commercial quantities of copper and iron that anyone gave serious thought to another canal. In 1852, the U.S. Government authorized "construction of a Ship Canal around the Falls of Sault Ste. Marie." On February 5, 1853, Michigan approved the canal project, but required a condition that the canal be completed by May 19, 1855 for the contract between the U.S. and the builder to be valid.

Sounds unusual? To the contrary, a great prize hung in the balance - three quarters of a million acres of Michigan land would be given to the builder of the canal in exchange for the canal. The canal was to be built and construction costs paid by private enterprise as a land grant project. On completion, the canal would be given to the United States, and the U.S., in turn, would give the builder 750,000 acres of federal lands in Michigan.

Lots of land and a canal - now this was a project Joseph Fellows and The Corning Company could understand.

In 1845, a Detroit attorney, James F. Joy, attracted the interest of Erastus Corning; and John W. Brooks, superintendent of the Auburn and Rochester railroad, in New York, was sent to inspect the Michigan Central Railroad as a possible purchase. Joy guided the purchase through the Michigan legislature. Money for the purchase also came from Bostonians John E. Thayer and John Murray Forbes. (The Michigan Central has often been described as a Boston business, but D.D. Williamson, a director of the Michigan Central, stated in 1848, "New York and Michigan [men] own a majority of the Road and can take the power in their own hands if they please." In the same light, Corning was recognized as one of the trustees of the eastern syndicate that developed the western railroad system we know today as the Burlington Route: the Chicago, Burlington & Quincy.)

As the federal, Michigan, and other state governments were passing appropriate legislation for a canal at Sault Ste. Marie, Joy was called on once again. This time, with Charles Thompson Harvey, a 22 year old salesman from the Fairbanks Scales Company, Joy guided the Corning and Fairbanks people to success. Saint Mary's Falls Ship Canal Company, organized on April 12, 1853, was given the contract to build the "Soo Canal." Corning was president of St. Mary's.

(Afterwards, Harvey settled permanently at Marquette, Michigan. His inventive engineering skills, developed in overseeing the Soo, led him to many other accomplishments. In 1874, he built a half mile of elevated track on Greenwich Street, in New York City, that became the New York "El.")

On June 18, 1855, the steamer *Illinois* locked through the Soo.

Success! Of the 750,000 acres that St. Mary's Company was entitled to receive in return, 563,000 acres were taken in timberland and 187,000 acres in land identified as mineral land. The later included 39,000 acres in Marquette County. The land grant was "sold" to Saint Mary's Canal Mineral Land Company (same people) - and yes, Joseph Fellows would eventually take title to his portion of the Michigan lands.

Soon, the Civil War, the War Between the States, would start.

Chapter 46

Not too many people knew of the close ties between these three men: William Winton, Horatio Pierce and John Law.

The War years (1861-1865) called for double involvement of the people of northeastern Pennsylvania. Not only did every community raise a company, or contribute to one at every call for men, but the need for supplies and fuel had everyone working overtime. Every county maintained a Roll of Honor, and after hostilities ended erected some type of monument to those who had fought and died. Local history books written thirty and forty years after the conflict would list in detail the names of every man who had served in the military, as if it were only yesterday that they had gone off to fight. No one was forgotten.

The War was brought home to the people of northeastern Pennsylvania in other ways. One such incident was a train wreck on the Erie R.R. A prisoner-of-war train was carrying captured Confederate soldiers to the newly opened Elmira Prison Camp (nicknamed Helmira) at Elmira, New York, on July 15, 1864, when it collided with a 50-car coal train near Shohola, in Pike County. Almost a 1,000 men, Confederate prisoners and Union guards, were involved in the collision, which was described as the great train wreck of the Civil War. Residents from both sides of the Delaware River rushed to render aid. (It was not until 1911 that all the dead were interred in Woodlawn National Cemetery at Elmira with a special marker, the Shohola Monument, memorializing the tragedy.)

At Hyde Park Borough, now a part of Scranton City, council members approved a "bounty" of $300, starting March 25, 1864, to be paid to each person "entering the army of the United States under any of the calls of the President, and who should be credited to the quota of Hyde Park." The double involvement for residents of northeastern Pennsylvania was taxing the people to the limits of their endurance.

On the other hand, there were some, like Ira Tripp, great-grandson of the Quaker Isaac Tripp who had first been spared by Indians and then later captured and killed by Indians, who found a definite personal advantage to going off to war. "During the later years of his life few Scrantonians were better known, or more respected, for their philanthropies than [Ira] Tripp. He had ample means, and was generous with them." But at the start of the hostilities, Tripp was on the verge of bankruptcy, having overextended himself financially. When his creditors learned that "their man" was involved in the defense of his country, they, equally patriotic, and "in deference to a custom for which the precedent was laid during the [1776] Revolution," delayed collection "until the close of the war." By that time, money saved and increased prices enabled Tripp to pay off his debt and keep his property, including his home (which had also been his great-grandfather's) and much of the coal land surrounding it.

And occasionally, "the call of the President for volunteers electrified the loyal men everywhere throughout this [Luzerne] as well as adjacent counties," and men such as Dr. Benjamin Throop, who "had not the slightest idea of going to the front" accompanied his friends of the Eighth Regiment as far as Harrisburg, along with Amos Meylert, to see them off to the war front. The next day, Throop, while "awaiting the departure of my train for home," was served with "the official document which made me surgeon of the Eighth Regiment, Pennsylvania Volunteers." (Meylert could not pass the physical.) Years later, Throop commented, "It was a service I gladly rendered then, and more fully appreciate the honor of now."

While many men went off to fight, the demands of war were forcing coal mines and iron furnaces to operate around the clock. Workers and their familes swelled population counts almost overnight. When the Scrantons first started operations in 1840, there were 100 people along the banks of the Lackawanna River near Roaring Brook Creek. Ten years later in 1850, the population was just under 3,000. This was certainly an impressive growth record for an area, but not overwhelming. Another ten years, and Scranton had tripled its population to 9,223. The war years ballooned population ever higher, and with the incorporation of Scranton as a city in 1866, bringing in Providence and Hyde Park Boroughs, the population count increased by 25,869 to a grand total of 35,092 in 1870! With so many new faces, all the necessities of living had to be accommodated in a hurry.

Coal, iron, and the railroads were the dominant industries. Money poured into the towns where these companies were headquartered. It wasn't long before Scranton gave evidence of growing at a faster rate than elsewhere.

To place a heavy fist of control as quickly as possible over the burgeoning community, the flow of money was the first attacked. (What better way to exercise control than through money?) Soon there would be banks galore, facilitated by changes in banking regulations. George W. Scranton, still very active with the DL&W R.R. and the LI&C Co., showed the Pennsylvania Legislator, Senator George Sanderson, the benefits of moving to Scranton when leaving office. Sanderson, elected from Towanda in Bradford County, took Scranton up on his offer, and shortly after the Meylerts opened their exchange and banking house in 1855, was in Scranton setting up his own banking business - George Sanderson & Company. The "& Company" stood for his brother-in-law Burton Kingsbury and his nephew George S. Kingsbury.

Sanderson & Co. prospered, eventually moving to larger quarters. "This was destroyed by fire in 1863, and a much handsomer structure arose on its ruins without interruption to business." The Scranton Bank that arose out of the ashes had as organizers W.W. Winton, Burton Kingsbury, George Sanderson, and George S. Kingsbury.

This was not William Winton's only involvement with banking. He would, in September 1863, receive a federal charter for the Second National Bank of Scranton, and would continue for a while the private banking house of Winton, Clark & Company, with Myron J. Clark, a merchant tailor in Scranton. As for the Second National, it was made up primarily of Isaac Dean (another of Winton's brothers-in-law), cousins through marriage Ira Tripp, Henry Griffin, and Joseph T. Fellows (a nephew of Joseph Fellows, Jr.), Elisha Phinney and Theodore F. Hunt. Phinney and Hunt would come to play very important parts in the life of Winton.

As for the First National Bank of Scranton, it received its charter on May 30, 1863. Those who had built and backed the DL&W and LI&C were the men who started the First National. Joseph H. Scranton assumed the presidency and J.J. Albright, vice-president. (Albright, the very same Joseph J. Albright who had gathered the first samples of iron ore at Slocum Hollow in 1836, had led a varied career since the time of his first visit to Scranton. In the interval he had also earned the nickname of "fire and water ironmaster," as every project he started had ended either by flooding or burning him out.)

The public image of the two great banks of Scranton, the First National and the Second National, was that they represented two opposing factions in the city, and were in constant rivalry with each other.

Meanwhile, Horatio S. Pierce made Carbondale his center of operations and was instrumental in starting the First National Bank of Carbondale by 1864. At the same time, Winton's and Pierce's friend as well as co-landowner with Pierce, John S. Law, headed up the Pittston Bank when it became the First National Bank of Pittston, also in 1864.

Not too many people knew of the close ties between these three men: William Winton, Horatio Pierce and John Law.

In a one year period, 1863-1864, Pierce, Law, and Winton had established banking institutions serving almost the entire northern anthracite coal field. The flow of money was theirs to direct. Remember, many of the large deposits going into their banks was their own money earned from coal royalties and coal sales. Where the money would go afterwards, only they might know.

Well, where did the money go?

It went everywhere! And this writer does mean everywhere.

For one place, it flowed out to the oil boom in Venango County, Pennsylvania. Not too many people knew that.

Winton and Pierce, with Law, were going to have to learn to live without Joseph Fellows. After all, in 1864, Joseph Fellows, Jr., was 82 years old. The Corning Company had disbanded at the end of its charter in 1855, and with its demise, the Corning Bank had also shut down, although not all the principals; Erastus Corning, Thomas Olcott and Fellows continued to be active when they felt the occasion called for it.

Hiram Bostwick had concerned Fellows for some time; Fellows worried that Bostwick was becoming too interested in pursuing a career of his own that might be in conflict with Corning Company interests. Fellows may have been right, because Bostwick began to plan in 1852 for a railroad from Lawrenceville, in Tioga County, across northern Pennsylvania through Potter County into McKean County. (The railroad was never built.)

The members of the Corning Company had adhered to their original plan to limit their formal association to 20 years. From the mid-1850s on, they concentrated on divvying up their assets, each according to his investment in the Company. There was no longer any need for the Company bank. Corning and Olcott both had their own banks in Albany. (There is no record of Erastus Corning ever visiting the community of Corning in Steuben County, New York.) Fellows, a much more complex man, always hugging the shadows (he never married), preferred to put his money about, here and there.

Too much money came in the war years, however, and another arrangement had to be found. The son of a lumberman, who had moved his family in the 1840s from the Hudson River to Tioga Township in Tioga County, Pennsylvania, went to Corning, New York following the death of his father. Quincy Winthrop Wellington caught Fellows' eye shortly after going to work as a clerk/bookkeeper in 1859 for one of the two banks which filled the void left by the demise of the Corning. What luck for Wellington! Three years later in 1862, the private banking house of Q.W. Wellington & Company opened its doors for business at Corning, New York. Wellington was doubly lucky (or was Fellows?), for his father-in-law had taken over control of Tioga County Bank, in Tioga Borough, near the end of 1859.

At the same time, younger men came forward to fill in some of the activities. After John Jay Phelps left Dundaff and Susquehanna County, with his bride Rachel Phinney, for New York City, brother Jaman Phelps moved directly to Scranton. Among his activities in the 1860s, he joined with William Winton to buy land that brother John Phelps owned in Scranton, for a development to be known as Phelps and Winton's Addition to Scranton. Brother Horace G. Phelps, on the other hand, moved to Corning, New York in the early 1840s, and under the watchful eye of Joseph Fellows, took an active part in the community there.

The Phelps brother-in-law Elisha Phinney also moved to Scranton. But because he was Gould Phinney's son, he had not left Dundaff, the town his father had founded, until 1856, several years after his father's death. Like Michael Meylert, Elisha Phinney helped George Scranton construct the Liggett's Gap/DL&W R.R.

From there, Phinney moved on to lease coal land from Winton south of Scranton, in today's Moosic Borough. After that, he joined Winton in the Second National Bank.

Two other men developed a close understanding with Winton. They were brothers Theodore F. Hunt and Alexander E. Hunt. And from Carbondale, Pierce listened to Eli E. Hendrick.

Chapter 47

Just as had Joseph Fellows, William W. Winton, Horatio S. Pierce and John S. Law used their money to underwrite further ventures.

Where did the money go?

Just as had Joseph Fellows, William W. Winton, Horatio S. Pierce and John S. Law used their money to underwrite further ventures.

One of the first ventures was helping Michael Meylert with his new county, Sullivan. For the first ten years after its founding in 1847, Sullivan had grown nicely, with an increase in population of about 2,000. But after that, there was almost no further growth, only a trickle of 500 more settlers over the following years. The new county seat, LaPorte, and all of Sullivan County, including Meylert's personal assets, were in jeopardy. No home-grown industry had started that could offer employment and sustain a population. So, out to LaPorte went Pierce with his able assistant James Stott, his co-landowner James Clarkson, his first mercantile partner in Carbondale Thomas Gillespie, Dickson Manufacturing

Company's James Dickson and others, to give Meylert a helping hand. LaPorte Tannery Company (later Thorne and McFarlane) was set up by 1860 .

For another investment closer to home, Winton and Ira Tripp helped Carter Edge Tool and Agricultural Implement Manufacturing Company get a start in 1864. Pulaski Carter had perfected the making of scythes, axes, and other tools. Carter Edge Tool Co., on Capouse Avenue in Scranton, went on to hold a prominent position among the city's industries for many years.

But it was the oil boom in western Pennsylvania that had the greatest attraction. Edwin L. Drake had discovered oil on August 27, 1859 at Titusville in Venango County, but the pressure of the War Between the States delayed further active exploration for the "black gold" for several years. What oil that was sent to market was put in barrels and rafted down the Allegheny River, like sending logs to market down the Delaware River. By 1864, the boom in real estate and oil wells was underway.

Horatio Pierce served as lead man for the oil project. Starting in November 1864, Pierce, with Daniel Devlin of New York City, began acquiring oil-rich properties in Venango County. Then for $50,000 Great Northern Oil Company purchased a half-interest in 68 acres in Cranberry Township, and for another $50,000, Great Northern purchased a half-interest in four and one-half acres in Franklin Borough, complete with refining equipment; both properties were acquired from Pierce and Devlin in January, 1865.

Great Northern Oil, by the way, was two-thirds owned by stock holders Horatio S. Pierce, John S. Law, Thomas Gillespie again, Charles Wurts, Joseph B. VanBergen (also of Carbondale and a Dickson in-law), and others. The great push to go ahead with developing oil property in Venango County was due to Eli E. Hendrick. Hendrick, at the time a resident of Franklin Borough (Venango County), had developed a technique for preparing oil for industrial use; he knew he could make money but he needed money to set it up. In Venango County, "the preparation of lubricating oil has met with greater success than refining. This is largely carried on at Franklin." Several firms including the "Hendrick Works have been very successful, preparing a lubricator adapted to the most delicate machinery, as well as to the common steam engine."

As the phrase would eventually develop - when "the boys from Scranton [Pennsylvania]" moved on a project, they did it the right way.

Between December of 1864 and August 1865, not only had Pierce purchased oil-rich property, and formed Great Northern Oil Company, but also helped to found Hendrick, Hurley & Company with Eli E. Hendrick, former U.S. Senator from Susquehanna County Galusha Grow, and Joseph VanBergen among others. And, in addition, with Amos N. Meylert, Isaac J. Post, and others, Pierce set up Latonia Town & Oil Company along the banks of the Allegheny River in Cranberry Township. (Members of Galusha Grow's family had been operating a store, Grow Brothers, in Carbondale for several years. Isaac J. Post was a great-nephew by marriage to William Jessup, his uncle Isaac L. Post having married Jessup's daughter Harriet.)

Concurrently with Pierce and Hendrick, on May 1, 1865 Elisha Phinney would bring together Isaac Post and his brother-in-law Alfred Hand (who had married Phoebe Ann, another of William Jessup's daughters), and Edward N. Willard, Theodore F. Hunt, Charles P. Mathews (Phinney's son-in-law), William Connell, and William Breck among others, to incorporate Lackawanna Petroleum Company. Lackawanna Petroleum operated 34 acres on Shafer Run, Venango County, and one-half of 140 acres in Clarion County.

All of these men formed a winning team.

About the same time the "boys from Scranton" were running around the oil fields of western Pennsylvania, Moses Taylor, Samuel Marsh, and John Blair were handing over (on August 1, 1864) their five-twelfths interest in more than 30,000 acres in Luzerne, Monroe and Wayne Counties to William E. Dodge. (Dodge already owned the other seven-twelfths.) Money was pouring into Scranton for everyone. The DL&W and LI&C were making profits; and among some of the original investors an agreement was reached. Dodge would buy out his fellow partners for $67,000. The trust created at the time of building DL&W track to the Delaware Water Gap had at last come to an end. And Dodge, all by himself, would manage 31,666 acres of timberland spread over Luzerne, Monroe and Wayne Counties.

Dodge put the headquarters for his lumber operation at Tobyhanna Mills, in Coolbaugh Township, Monroe County. The DL&W swept through the center of the little community on its way to Stroudsburg. While the Dodge name was on the land, and everyone knew Dodge was involved with the railroad, the timbering and saw mill were a joint venture - Dodge, Meggs & Company. (The town of Tobyhanna and the Tobyhanna Army Depot stand where Dodge, Meggs & Co. once operated.)

Among the men who worked for the company, Charles Hebard was anxious to advance himself. As Dodge worked timberland across Pennsylvania, especially with the Pennsylvania Joint Land and Lumber Company, Hebard hoped to make some money for himself. At the beginning of the 1870s, he organized a saw mill near Williamsport (Lycoming County) with two others, under the name of Hebard, Foresman and Smith. Hebard's future, however, would lead him elsewhere.

Some of the money Pierce and Winton were accumulating stayed around Scranton, too. In a little over a year's time, from March 1865 to May 1866, a stove manufacturing company and a public transportation system were established. The People's Street Railway Company of Luzerne County (formed March 23, 1865) began with William Winton, brother-in-law Abram Dunning and relatives-by-marriage Ira Tripp and Joseph T. Fellows, with others, calling for the incorporation of a public transportation/trolley system between the heart of Scranton City and the Providence section. James Blair, brother of John I. Blair, was named president of the trolley line, and Alfred Hand (Jessup's son-in-law) secretary, Henry B. Rockwell (of Winton's New York & Pennsylvania Coal Co.) and Theodore F. Hunt joined the endeavor.

A year later, on March 7, 1866, George Sanderson, with George S. Kingsbury, William Breck, William Monies, and Jacob Robinson became the officers of the Scranton & Providence Passenger Railway Company. This trolley line connected Scranton with Providence, but by a slightly different route than the People's road. How separate were the two companies? Remember, Winton, Sanderson and Kingsbury had gone together to incorporate Sanderson's bank at one time, and Breck had been a part of Phinney's efforts on Lackawanna Petroleum. Regardless, the two trolley lines did not stay apart for very long. "A lively contest over the right-of-way was settled by Judge Conyngham, and on the 11th of May, 1868, both roads, merged into one, [as] People's Street Railway Company, were opened to the public."

Winton continued as a director of the merged trolley lines through several changes, almost resembling a game of musical chairs among friends. Theodore Hunt stepped aside so that William Connell could sit on the board; Joseph H. Scranton became a director, followed by his son William W. Scranton; Ira Tripp left so that George Sanderson would have a board seat; Sanderson, in turn, left so that George H. Catlin, a younger man, could serve on the board. (By 1888, the Board of Directors would have a majority of New York City investors and - guess who? - William W. Winton and William H. Jessup, son of William Jessup!)

The stove manufacturing company, known as The Scranton Stove and Manufacturing Company, was very important. One of the successes of anthracite had been its desirability as a home heating and cooking fuel. Patented stoves meant more coal sales. The Scranton Stove was an 1866 reorganization of an existing business. This time Pierce took charge; both Hunts (Theodore and Alexander), Sanford Grant, and John A. Price formed a part of the board. Winton would show up later as president for a while.

And then there was the Scranton Board of Trade, forerunner of the city's Chamber of Commerce. Again Elisha Phinney stepped forward to get the ball rolling on December 12, 1867. It wasn't until 1870 that Winton became secretary.

When the War Between the States, the Civil War, came to an end, Joseph Fellows was 83 years old. He was still the Pulteney agent. In addition, due to the formal dissolution of the Corning Company, Fellows had his own vast real estate interests to manage. Readjusting to Fellows' advanced years, Horace G. Phelps, who had left Dundaff (Susquehanna County, Pennsylvania) to take an active role in the municipal affairs of Corning under Fellows, and who had been one of the three petitioners asking the court to approve the 1848 incorporation of Corning village, left Corning to set up a business and be active in Scranton. At the same time (in 1865), Fellows' nephew John Heermans and his second wife Nancy left Scranton to settle at Corning. There, Heermans assumed the duties of private secretary to his uncle. (Heermans had gone to West Virginia in the 1850s for Fellows.)

With John Heermans to help him, Fellows turned to concentrate on the future of the community he had been so active in starting. The first days of Corning, New York, had been mainly as a terminus for the Blossburg coal brought out of Pennsylvania; the coal then spawned a secondary industry - timbering. Clear cutting was then not only a way of life, but an acceptable philosophy. By clear cutting, the land was prepared for newly arrived immigrant settlers to plant crops quickly and establish farms. The second growth of trees that would come along would be different from what had been cut down, and would be beneficial to the soil. But of course, with clear cut, the lumber company is cutting away at its own future.

The junction of the Tioga, Conhocton, and Canisteo Rivers was ideal for transhipping a heavy product, such as coal. It also proved to be an ideal location for saw mills and a lumber company.

Fox, Weston & Bronson took over the sawmill at Painted Post/Gang Mills, just over the boundary from Pulteney lands and Corning (but still in Steuben County) in 1846. The lumber company grew "until it has become the most extensive lumber establishment and firm in town, county or State, and the nucleus of perhaps the most extensive lumbering business in the United States." It processed over 13 million board feet in 1869 alone. Many of the logs Fox Weston processed came from Tioga County, Pennsylvania. But under the accepted practice of the times, which was to clear cut, the company would eventually exhaust its supply of timber. So for all the activity at the mill, in another six years the move to shut down operations at Gang Mills had begun; and in 1880, Fox Weston closed its doors at Painted Post/Gang Mills.

Joseph Fellows was well aware in 1860 that the day would soon come when there would be no more trees to cut. The area would have to have another major business to fall back on. Events in Wayne County, Pennsylvania, at a place called White Mills, halfway between Honesdale and Hawley, caught his eye. In 1861, President and Mrs. Abraham Lincoln had ordered cut glass tableware from a Brooklyn, New York manufacturer, C. Dorflinger & Company, through a Washington, D.C. retail store. The following year, in the summer of 1862, manufacturer Christian Dorflinger, owner of the glass factories (he had more than one), had taken a leave of absence because of ill health, and retired to White Mills to recuperate. By 1867, Dorflinger had decided to re-establish his glass works at White Mills, with the D&H Canal only a few feet from where he would erect a glass house and five-pot furnace. From White Mills, via the D&H Canal, beautiful and elegant cut glass pieces would be shipped more than once to the White House.

Dorflinger, with his Concord Street Flint Glass Works, Long Island Flint Glass Works and Greenpoint Flint Glass Works, was not the only glass blower operating in Brooklyn. Amory Houghton, a glass manufacturer from Somerville, Massachusetts had purchased the Brooklyn Flint Glass Company in 1864. By 1868, Houghton was ready to emulate Dorflinger and move out of Brooklyn. Houghton, however, would respond to the persuasions of Corning, New York and take his glass works there. Joseph Fellows, Erastus Corning, and Thomas Olcott, the driving forces behind the Corning Company, came together again. Each man contributed $10,000 to help set up Houghton as the Corning Flint Glass Works. Later, the company became known as the Corning Glass Works; and by December 1870 a $1,000 order of glass, cut by Hoare & Dailey who had also re-established at Corning, New York, was on its way to another U.S. President, Ulysses S. Grant. (By 1881, Corning Glass was making glass bulbs for Thomas Edison's latest invention, the electric light bulb. And by 1977, over a century later, 30,000 people would be employed by Corning Glass.)

Meanwhile, back in Pennsylvania, about a month before Corning Glass was to begin its manufacturing operations in Steuben County, New York, The Honorable William Jessup died at Montrose, Susquehanna County, on September 11, 1868.

Chapter 48

With money moving in and out of private accounts presided over by Pierce and Winton - who would know? Yes, indeed, who would know what was going on?

And then, the most powerful of banks was formed - Scranton Trust Company and Savings Bank.

As the year 1870 approached, the power of the railroads and anthracite coal solidified. As to who needed whom, the question, like "which came first, the chicken or the egg?" could never be resolved. Without the railroads, coal would never make it to the consumer. Without coal, railroads and steam locomotives didn't operate too well, if at all. The demand for anthracite had increased to the point where each of the last four years of the 1860s had seen over 12 million tons mined and shipped to market. In 1870, 16,182,191 tons would be sold. And this in an industry which had shipped just 365 tons in 1820. By 1870, the industry was a $60 million business, and was the foundation for a multitude of other industries and businesses. The financial world of stocks and investments was one such enterprise. Buying and selling huge blocks of railroad stocks could make a fortune for the lucky individual, without his ever going near a train station, much less descend into a mine.

The great and famous financiers of Wall Street in New York City would willingly buy up government bonds and treasury notes when issued, because industry and its taxes were behind the governments' needs for money. Underwriters were no longer faced with the problem of having to market securities abroad; buyers could be found in the United States.

The industrial world's wheels would turn only if they had power. In the nineteenth century, the best available source for power was coal, especially for steam engines, both stationary and moving. At the heart of the energy supply lay the coal operator and the railroad that would carry away the mined coal.

The Scranton Trust & Savings Bank was chartered in 1869, but did not open its doors until 1870. And my, my - look who's there! Horatio S. Pierce is president. Thomas Dickson is vice-president. William W. Winton is treasurer. Horace G. Phelps is secretary. And who has joined these gentlemen on the Board of Directors? Joseph H. Scranton, John Brisbin, and Ira Tripp. (Yes, it is the same Horace Phelps. As for the others: Scranton had helped metamorphize Scrantons & Grant into LI&C, and is still president of the First National and Scranton Gas & Water; Dickson is now president of D&H while the rest of his family continues to operate Dickson Manufacturing; and Brisbin has just stepped aside as president of DL&W to serve as general counsel and consultant to the railroad.) What a bank! Who would know what these men, individually and/or collectively were up to? For some of these men, it was characteristic to attend personally to business matters; to limit use of the telegraph, available since the early 1850s, to brief, prosaic messages such as 'Your order was shipped today'; and to rely on confidential agents when initiating new courses of action.

With money moving in and out of private accounts presided over by Pierce and Winton - who would know? Yes, indeed, who would know what was going on?

Just as Scranton Trust and Savings was forming, Winton himself had joined with Edward Dolph, a cousin on his wife's side of the family, and Nathaniel Halstead to buy land surrounding Pierce's Grassy Island Coal and R.R. property at Jessup. In all, between 1869 and 1872, these three men would acquire six warrant tracts, totaling over 2,000 acres, all very rich in coal.

The Mid-Valley (halfway between Scranton and Carbondale) had never been substantially developed, as far as coal mining went. Pierce, in charge of Grassy Island, and with co-owners William Monies, Lewis Pughe, Edward N. Willard and Clark E. Rogers (who had bought Hollenback's portion) would lease mineral rights from time to time. One such lease in 1870 was to Edward Jones. Jones, a year later in 1871, assigned his lease to Glenwood Coal Company. Glenwood was already mining in the Carbondale area. The president of Glenwood Coal Co. was Jay Gould.

Gould had entered the national limelight when he and two others, James Fisk and Daniel Drew, had cornered the gold market in 1869. In the financial panic that resulted, the three men made a profit of about $11 million; President U.S. Grant almost lost his White House residency; and the country went into one of the worst depressions in its history, starting on "Black Friday" in October 1869. Gould's rise to financial dominance had started in the 1850s with a tannery in Gouldsboro, Wayne County, near the headwaters of the Lehigh River.

Generally speaking, mining in the Mid-Valley before the 1870s had been pursued in a desultory manner. Lewis S. Watres, who had supplied lumber to Jessup and Meylert's Lackawanna Rail Road, opened up a small mine and worked it until its vein gave out. He named the place where he was living Mt. Vernon, in an area sandwiched between Jessup and Archbald. But when Dolph, Winton and Halstead took over and began expanding operations, one of the first places to change was Mt. Vernon. The small mining community was renamed Winton, and was ready to open for work in 1873. The change that came about at Mt. Vernon/Winton was due primarily to the capabilities of one particular mining engineer, George Filer. While John Jermyn had led the New York and Pennsylvania Coal Co. to profits and success, Filer, who had come to the Scranton area from Pottsville in 1849 to open up Pennsylvania Coal Co. mining properties, was the one who demonstrated an almost uncanny capability for locating highly profitable veins of anthracite where others believed all coal had been exhausted. Filer found that there was still abundant coal at Mt. Vernon. On the strength of his appraisal, Winton, Dolph and Halstead spent three years amassing the acreage necessary for a large scale mining operation, especially when joined with Pierce's Grassy Island.

But while Winton and the others were setting up coal operations in the same area where Jessup and Meylert had once operated, long-established coal companies throughout all the anthracite coal fields were forming a closed group. Spoken of as "the combination," the group's existence was "for the purpose of monopolizing trade."

Thomas Dickson attributed "the combination" to the aftereffects of the War Between the States, the Civil War, when manufacturers had not yet geared down to peacetime production. The supply of coal on hand was far in excess of demand, and in 1872 everybody in the coal industry was losing money. The leader in forming "the combination" was David Gowen, head of the Philadelphia & Reading Railroad Company (Reading R.R.), who had been able to project the Reading into a dominant position in the Schuylkill County coal field area. But while the original motivation may have been a strong sense for survival, in reality "the combination" came to represent "merciless taxation." Coal consumer and coal miner alike suffered under high prices and days without work and without pay. The coal companies also had their own problems, too. In regulating the output from the mines, "the combination" established a kind of "pecking order." Reading R.R. was designated to produce about 38% of all coal. In line following after, Lehigh Coal & Navigation Co. was allowed about 20% of the market, the DL&W - about 16%, D&H - 11%, Pennsylvania R.R. (headquartered at Philadelphia) - 8%, Pennsylvania Coal Co. - 5% (and in 1885, New York, Lake Erie & Western Co. - 1.5% of the anthracite coal market). Before "the combination" was in control, a ton of coal cost $3.70 to bring into New York City. Afterwards in 1873, coal was $4.90, a year later $5.13/ton. "In the first ten years of the existence of the coal combination the cost of the monopoly to consumers of anthracite was estimated to be not less than $250 million."

Were the activities of "the combination" a trust? Was this what the Sherman Antitrust (in 1890) and Clayton (1914) Acts were enacted to stop?

What about the arrangement that had existed at Lowell, Massachusetts since 1823? At Lowell, all the manufacturing companies kept an eye on each other. If one lowered pay rates or increased prices, the other companies followed. At Lowell, it was a generally recognized fact that managers of the different mills would meet in times of economic crises to spell out mutually agreeble plans of action. So-called "interlocking directorates" (the same individual(s) serving on the boards of directors of more than one company in like businesses) were an acceptable way of life at Lowell. In fact, the interlocking directorates were preferred, because it meant no one company would engage in activities to the detriment of the other companies located along the Merrimack River.

Why were monopolistic trust actions condoned at Lowell? Because the factories all shared the same source of energy - water power; a business could operate only if it held a water power right. (All factories were hooked up to the same waterwheel.) But however valid the reason, the bottom line result was that commercial competition was stifled at Lowell.

But cloth manufacturing was not limited to Lowell. In a setup such as at Lowell, cloth manufacturing was limited only to the extent that there were fastflowing streams/rivers. "The combination" that Gowen of the Reading instituted for anthracite coal had far more effect as long as anthracite was the only coal used. Some of the owners of anthracite coal lands, and some of the owners of anthracite-carrying railroads, reacted to "the combination" and the control it was exercising over their right to profits. They looked elsewhere for coal and built their own railroad.

Chapter 49

And what would be the fate of anthracite land owners and railroad carriers in the northern coal field at Scranton?

Joseph Fellows, Jr. retired as Pulteney land agent in 1871. He was 89 years old. He died two years later in 1873 at the age of 91.

Almost at the same time, a much younger man, Erastus Corning, died in 1872 at the age of 78.

It was the end of an era.

Fellows' death coincided with the growth of the coal combination. "By restricting production and by advancing prices, it ('the combination') had crippled vast iron interests, decreasing the demand for anthracite coal in the reduction of ores, forcing iron furnaces out of blast, and placing nearly all the industries of Eastern Pennsylvania at its mercy." Yes, that's true, the great iron mills along the Lehigh Valley, Bethlehem and Easton, who had carried the North through The War to victory, were fast becoming a thing of the past.

Does it seem unbelievable, the mischief this "combination" is effecting? How will the coal companies and railroads of the lower coal fields survive, let alone those in the upper coal fields? Hasn't Gowen and the Reading R.R. actually snuffed out their own existence? Ha! Guess again.

"The great anthracite coal companies had prepared for a change." The companies were the Reading R.R., Lehigh Valley R.R., and Pennsylvania R.R., and they had already "purchased immense tracts of bituminous coal fields" in "Central and Western Pennsylvania, Virginia, and Maryland." The shift to Pittsburgh as the center of iron and steel making in the United States had begun. It was expected that the opening of the new, bituminous (soft coal) fields would affect the anthracite coal trade. "There seemed no hope for the increased use of anthracite" except for domestic purposes.

And what would be the fate of anthracite land owners and railroad carriers in the northern coal field at Scranton?

Joseph Fellows had sent his nephew Sylvanus Heermans to Preston County, West Virginia in 1848. The bituminous coal fields were not unknown to "the boys from Scranton."

With William Winton's support, Edward Dolph left the emerging mining operations and village at Winton, in the Mid-Valley, and traveled to West Virginia. There, by the end of April 1873, Dolph, for himself and Winton, had purchased 21,000 acres in Pocohontas and Randolph Counties, along the Cheat and Greenbrier Rivers.

So much for the Reading and Pennsy (Pennsylvania R.R.) and their ideas for squeezing everybody else out of the coal business.

And at about the same time, in February 1873, a group from Wilkes-Barre, consisting of John B. Wood, A.J. Davis, C.P. Kidder, J.D. Hoyt, and Sheldon Reynolds, acquired one square mile of land at Silver City, Grant County, in the Territory of New Mexico. To the north, Thomas Dickson carried $300,000 of Scranton money towards the purchase of 23,000 acres rich in iron ore in northern New York State.

But that wasn't all; Dolph was on his way to the Upper Peninsula of Michigan by October of the same year (1873). There, in Ontonagon County on the shores of Lake Superior, Dolph formed the Scranton Silver Mining Company. This time Dolph was not alone. Charles Hebard would be a part of the company. Hebard, who knew his wood, who had tried to get his own lumber and saw mill company going while working for William E. Dodge at Tobyhanna Mills in the 1860s, was now a part of the team, "the boys from Scranton." (Dodge would own timberland in the Upper Peninsula. Most of it was to the east of Marquette, in Alger County.)

The Scranton Silver Mining Company, and later the Luzerne Silver Mining Company, took title to just 165 acres in Ontonagon County. Luzerne Company included Dolph, Oscar C. Moore, Edrick Frear, and A. C. Sisson from the Scranton area. The Scranton Silver Mining Company, except for Dolph, was composed of an entirely different group of men. Besides Hebard, Nathaniel Halsted, working with Dolph and Winton to develop the coal at Winton and Jessup, was also a part of the Scranton Company mining at Ontonagon. So too were Henry C. Thurber, John Spalding, and Alfred P. Swineford, from Marquette, Michigan. Elisha Phinney showed up as president of Scranton Silver in 1874, and William Winton as treasurer in 1875.

Hebard's mission in coming to the Upper Peninsula, however, was to evaluate the area for timber purposes. What Hebard saw were white pine forests stretching for miles. With Thurber, Hebard proceeded to log the southern shore of Lake Superior from Ontonagon to Marquette, including all of Keweenaw Point. The two men engaged John Longyear as landlooker. Hebard's association with "the boys from Scranton" continued. Of the others who helped form Scranton Silver Mining, Swineford was already editor and publisher of the *Marquette Mining Journal* when he joined the Scranton Company. Swineford went on to become the first resident governor of the District of Alaska in 1885. He and his family lived in Alaska for 25 years before returning to Marquette.

Samuel Sloan also showed up in Marquette, Michigan. Sloan had recently become president of the DL&W at John Brisbin's retirement, and had proven himself a very able railroad executive. He had served as president of the Hudson River R.R. for ten years, ending in 1865. In 1867, he was elected president of the Scranton-based DL&W railroad, having spent the two year interval as railroad commissioner for trunk lines.

Sloan was in the Upper Peninsula for one reason: The Houghton & Ontonagon Railroad Company and the Marquette & Ontonagon Railroad Company had just merged, becoming The Marquette, Houghton & Ontonagon Railroad Company; Sloan would be president of the newly created MH&O R.R. in addition to his presidential duties with the DL&W. (Sloan left the MH&O before 1885 and the railroad was acquired by Duluth South Shore & Atlantic Railway Company in 1890. Duluth South Shore became a part of the Soo Line in 1961.)

While Sloan was working on railroad logistics in the Upper Peninsula, the DL&W, along with the D&H and Pennsylvania Coal Co., remained trapped in coal production quotas that offered no future. The "boys from Scranton" had some ideas on that, too, albeit of the "band-aid" variety.

On December 30, 1873, Theodore F. Hunt, along with his brother Alexander E. Hunt and New Yorker William H. Meeker, entered into an agreement with the DL&W (Yes, signed by Samuel Sloan and DL&W Secretary Fred F. Chambers) to construct a railroad line from Winton, Pennsylvania, to meet the main line of the DL&W (at Scranton). The Hunts and Meeker would build the track, using the old roadbed of the Lackawanna Rail Road, to specifications laid out by the DL&W and subject to DL&W supervision. Why were the Hunts and Meeker doing this? They had just entered into an agreement with George Filer and Thomas Livey (William Winton's son-in-law) to buy all the coal Filer and Livey would be mining at Winton/Jessup. The Hunts and Meeker needed assurance that they could get their coal to tidewater and market.

The Hunts and Meeker also promised the DL&W that they would ship no more than 60,000 tons in 1874, 220,000 tons during 1875, and 300,000 tons yearly starting with 1876, with the right to ship an additional 200,000 tons after giving notice to DL&W.

So much for the "the combination" and percentage quotas.

The approximate 12 mile length of track would become known as the Winton Branch of the DL&W.

(Railroad and DL&W expert and historian Thomas T. Taber, III commented on the developments at Winton following the Hunts, Meeker contract with the DL&W, in his 1987 publication *Railroads of Pennsylvania Encyclopedia and Atlas*: "The largest number of line haul railroads in a small area, each serving its own customers, was at Winton, north of Scranton. Five different railroads could be found with branch lines within a 3/8 mile radius - Lackawanna, Ontario and Western; Delaware and Hudson; Erie; New York Susquehanna & Western. Nowhere else in the entire United States were there so many different roads each running local freights. All served coal mines.")

Chapter 50

Governor John F. Hartranft proclaimed Lackawanna County, the 67th county in the Commonwealth of Pennsylvania. Scranton was designated the county seat.

By the 1870s, land ownership was no longer the exclusive sign of wealth and power. Wealth, and its accompanying power, could now be obtained from a plethora of sources, thanks to the steam engine and advances in technology. The Industrial Revolution was in full swing; and with it, the universally recognizable outward sign of wealth came to be personal aggrandizement. Elegant life styles, elegant clothing, elegant personal property and living accommodations were displayed with abandon. It was then that those "who had not" suddenly realized they would in all probability never have what others had, regardless of freedom, liberty, and democracy - the equality of man. Unfortunately, a few of the religious leaders at this time, especially those who ministered to the very rich, decided to soothe away any conscience the sudden new rich might have about their excess piles of money, by declaring that wealth was a sign of approval from God. This, of course, was a carte blanche invitation to keep on accumulating even more wealth as fast as possible .

Now openly declared to be under the disapproving eye of God, what was left to the good, honest men and law-abiding workers? Very little but the strike and the ballot box remained to them. In July and August of 1877, rioters erupted everywhere there was industralization. Much of Pennsylvania Railroad's buildings, equipment and track at Pittsburgh was destroyed. The same was true for railroads and industries in Chicago, Baltimore and other cities.

The northern anthracite coal field, particularly from Carbondale to Scranton, had been among the first areas to unionize - under Terence Powderly and the Knights of Labor. But remember, the old Connecticut claim and the people who settled the Wyoming Valley were a shadow of the past that could not be wiped away overnight, any more than the belief in land as the only form of wealth had been immediately cast off after the American Revolution. Arrogant displays of power by wealth were rare displays among the plain and simple folk of Quaker and Moravian background.

At Scranton, the rioting lasted only a few hours; almost no damage was done to mercantile property. There was a killing, and a trial followed. After that, it was the ballot box that was used to make changes, Powderly leading the way. Powderly was so popular and so successful that he was elected mayor of Scranton the following year (1878).

The year before the rioting, in 1876, William Winton's oldest son Aretus Heermans Winton, called A.H., knew what he wanted more than anything else. A.H. had become a lawyer. He was an integral part of the family's involvement, from Corning, New York, to the Upper Peninsula of Michigan, to West Virginia, to New Mexico, to wherever else the family had an interest. A.H. wanted something more than what his family could give him.

What A.H. wanted, he wanted first for himself, then for all his extended family members and for his neighbors. His neighbors can best be described today as comprising everyone living north of the Wilkes-Barre-Scranton Airport to near the headwaters of the Lackawanna River above Carbondale. A.H. Winton wanted to live in his own county; away with Wilkes-Barre and Luzerne County!

Residents of the Lackawanna River valley had been trying to live under their own jurisdiction since the day Joseph Griffin was sent to Harrisburg as a legislator in 1837. William Merrifield, one of the first to settle in Hyde Park, kept up the pressure for a new county during his term at Harrisburg in 1844. When Abram Dunning was elected in 1852, the push for separate status was renewed with vigor. But everyone's efforts had always ended in nothing. All the industry and mining, up to Carbondale, remained a part of Luzerne County's tax base. A.H. Winton promised to renew the effort and this time to succeed.

It was a colossal struggle, played out against labor's unrest. The courthouse crowd at Wilkes-Barre did not want to lose the upper half of Luzerne County. At one point, A.H. Winton had had all the steps in place necessary to dismember Scranton, thereby changing it from a third class city back to a borough. He was going to take the reestablished Boroughs of Providence and Hyde Park with him, and with the newly created Borough of Winton, make a new county.

When the time came to vote for new county status, labor and commerce came to the polls en masse, 9,615 strong. The struggle ended in victory for the new county proponents at noon time on August 21, 1878, when

Governor John F. Hartranft proclaimed Lackawanna County, the 67th county in the Commonwealth of Pennsylvania. Scranton was designated the county seat.

Later, after the ceremonial laying of the cornerstone for the Lackawanna County Courthouse on May 25, 1882, there was a gala banquet for the "friends of the new county movement" at the

Wyoming House at the corner of Lackawanna and Wyoming Avenues (the hotel, erected by George Scranton in 1852, eventually became a department store, and is presently the temporary headquarters for the banking firm National Westminster.)

Edward Merrifield, son of William Merrifield, addressed the happy group gathered at the Wyoming House that evening long ago. In part, Merrifield said, "I come here tonight to speak the truth of history; I am here to give credit to the men who got Lackawanna County. Nobody responded to this call but W. [William] W. Winton, H. [Horatio] S. Pierce, B. [Benjamin] H. Throop, George Sanderson, E. [Edward] N. Willard, A.H. Winton - and I was there, of course."

Names sound familiar?

A stenographic record was made of the speeches that evening. Robert H. McKune, who had been mayor of Scranton in 1877 but lost the office in 1878 to labor leader Terence V. Powderly, compiled the record and a brief history into a bound book. The reason, McKune wrote in the book's preface, "is that the events and incidents that are here narrated I believe are worthy to be gathered together; so that those who shall come after us may have some appreciation of how joyous we were...it was our fortune to see our fondest wishes fully realized, and we made happy, on the evening of August 13th, 1878." (The voting on the issue took place on Tuesday, August 13, but the Governor did not proclaim the county until August 21.)

The new county status for the upper end of Luzerne County, however, had not been gained without a sacrifice. There was no more Second National Bank of Scranton. In order to create Lackawanna County, the Wintons had had to give up their bank. Pierce, with his First National of Carbondale, and John S. Law, with his First National of Pittston in Luzerne County, helped as much as they could. But they had to be careful so as not to get caught and also sucked under to failure.

So William Winton and his easily identified activities were subjected to attacks - just at the time that Winton, with Pierce and Law, was about to embark on another venture.

In July of 1875, the three men had publicly acknowledged their close relationship to each other when they joined together to purchase 180 acres of coal land in Blakely Township for $42,626.61. (The Blakely Township of the 1870s was far reduced from the Blakely of the 1820s, when it comprised all the land between the original Providence Township and Carbondale. Over the intervening years, all of the mid-valley communities were carved out of Blakely.) The Pierce, Law, and Winton tract was on the west side of the Lackawanna River and a little south of Winton and Jessup.

The tract in Blakely was rich in coal. But more significantly, by openly linking their names together, these men demonstrated that the railroads were not the only owners of large amounts of coal land. And for all their coal acres from Pittston to Carbondale, Pierce, Law and Winton were not members of "the combination."

Chapter 51

On June 10, 1881, The New York, Susquehanna and Western Railroad Company (NYS&W R.R.) was formed. It was to be a continuous track from the Pennsylvania coal region to tidewater at Jersey City, New Jersey.

The closing of the Second National Bank of Scranton had become a very serious affair. As soon as the bank's doors had closed, a federal bank examiner was called in to examine the books. The examiner's initial pronouncement, that the problem was one of cash flow and that the assets of the bank more than exceeded its debts and liabilities, never changed. William Winton himself, oozing the confidence of a sound financial institution executive, let it be known the bank would reopen as soon as adjustments were made. Unfortunately, Winton's expectations were not what the now truncated Luzerne County aficionados wanted to hear. If A.H. Winton's cry had been "Away with Luzerne," then the whispered shout echoing back was "Away with the Wintons."

Where would it all end? How would it end? For one, the existence of Lackawanna County was a fait accompli; that would not change. For protection, where there was uncertainty, the "boys from Scranton" reduced exposure as much as possible. On April 12, 1879, the Scranton Trust Company and Savings Bank closed down voluntarily, never to reopen. "It was resolved that it is expedient to wind up the affairs of this bank," determined the stockholders. And how did they decide to wind up affairs? "At a meeting of the Directors of the said corporation it was resolved that a general assignment of all the estate real and personal of this bank be made to Horatio S. Pierce."

Where there was no uncertainty, only a path marked "real trouble" ahead, A. H. Winton gave counsel. Catherine Heermans Winton, William Winton's wife and A.H. Winton's mother, came forward - with judgment notes executed by her husband to herself, going back 22 years to the days of New York and Pennsylvania Coal Company. The Wintons, father and sons, had used Catherine's inheritance from her father, Henry Heermans, to help set up the operation. Catherine Winton asked the Sheriff of Lackawanna County to sell William Winton's assets, as she wanted her money back. (Winton's own assets - no assets of the Second National were involved.)

The sheriff's sale was scheduled for December 1879; the sale took place. Winton's properties were sold to the highest and best bidder, Catherine H. Winton. Then the meleé started. Winton antagonists obtained an injunction - not to stop the sale, but to stop the recording of the deeds to Catherine Winton. Almost 4,000 acres of coal land in Lackawanna County were put in limbo. And there the matter has stood ever since.

But William Winton, Horatio Pierce and John S. Law had been making big plans when they got caught up in the euphoria of erecting Lackawanna County. The plans may have been put on temporary hold, but they certainly were not to be forgotten. So how to handle the situation? Winton made sure all the depositors were paid, that they received all of their money back. The return of the money was made in four payments by selling Second National's assets over a seven year period, all the while Winton holding to a vision of reopening. But it seemed that every time things quieted down and William Winton thought the worst was over, an angry depositor would come forward with a lawsuit claiming whatever; that the bank examiner wasn't doing things correctly; that there was fraud; that money was not being reimbursed quickly enough, etc. Finally, a receiver was appointed, and the liquidation of the assets took even longer.

(Nathaniel Halstead with Winton and Edward Dolph, in the Mid-Valley and at Ontonagon County, Michigan, had borrowed from the Second National. In winding up the bank's affairs, the Receiver allowed the Wintons personally to cover some of Halstead's debts. Halstead surrendered some of his interests to the Wintons.)

In the meantime, Pierce, Law and Winton had to move ahead with their own plans. "At the close of the war [1865] anthracite producing lands were in the hands of individuals, but in 1867 the railroads began to acquire possession of them. This acquisition went steadily forward, and at the end of 1876 the Delaware, Lackawanna & Western Railroad Company, the Delaware & Hudson Canal Company, the Lehigh Valley Railroad Company, the Central Railroad Company of New Jersey, the Pennsylvania Coal Company, and the Philadelphia & Reading Railroad Company had acquired by purchase or lease the greater portion of anthracite producing territory." Add to that the Pennsylvania R.R., Erie, and Lehigh Coal & Navigation, and the fact that the Reading had absorbed Jersey Central and you have "the combination." And "the combination," headed by David Gowen of the Reading, was dictating who should mine how much coal and who should make how much money.

The northern anthracite coal field was getting a very short shift from Gowen and the lower coal field under the terms laid down by "the combination." All three northern companies, DL&W, D&H, and Pennsylvania Coal together, could not put out as much coal as just the Reading - a total of 32% of the anthracite trade to Reading's 38%. This was not a future for businessmen.

Henry VIII, King of England, had faced the same kind of situation in the 1500s. His little island kingdom had been allotted a certain place in the world - behind Spain, behind Portugal and after the rest of Europe. The Pope at Rome, as arbitrator, had divided the world in half by the Treaty of Tordesillas in 1494, to stave off rivalry between "the two most Christian countries" - Spain and Portugal. Except for a small eastern tip of Brazil, Spain got all of the Americas, and the incredible wealth of South America.

With Spain pouring new supplies of gold and silver into her coffers with every ship's arrival, European economy became so skewed in favor of Spain that the royal monarchs of lesser fiefdoms could never hope to rise out of debtor nation status. As the head of a Christian country, Henry VIII could never have ignored a treaty with a papal seal. Only after he broke with Rome and the Pope over annulment and remarriage in 1531, was he free to disregard the Tordesillas treaty and send his sailors around the world to explore and claim lands in the so-called Spanish and Portuguese territories, thus laying the foundation for "the British Empire on which the sun never sets."

So, assuming that there would be no end to "the combination", and that the railroads of the northern anthracite coal field would continue to be bound to abide by "the combination's" edict, Pierce, Law and Winton prepared to bring another, new railroad into the northern coal field.

As early as 1836, John I. Blair had been trying to get a railroad into the anthracite region. Blair lived in New Jersey, about ten miles due east of the Delaware Water Gap; the community was known as Gravel Hill at the time. (Before Blair died in 1899, he had earned the sobriquet "Railroad King of the West" and was worth over $60 million. Gravel Hill changed its name to Blairstown in 1839.)

In 1836, the enterprising Blair wanted to construct a railroad from the anthracite region across New Jersey to the New York City area. He was a member of a group that obtained a New Jersey charter to build the New Jersey, Hudson and Delaware Railroad Company (NJH&D). While the NJH&D was still in the paper stage, Blair became a Special Partner in the Scrantons' activities in Providence Township on the banks of the Lackawanna River and Roaring Brook Creek, and shortly thereafter a major backer of the Delaware, Lackawanna & Western Railroad (DL&W). The DL&W, by an agreement in 1854, would use The Central Railroad of New Jersey to traverse that state. (Not so surprising, co-partner William E. Dodge, with his partnership interests in DL&W, Lackawanna Iron & Coal Co. (LI&C), and Pennsylvania timberlands, was a member of The Central's board of directors.) As the DL&W became a reality, the pressing desire by Blair to develop NJH&D faded into oblivion.

The DL&W, as a corporate entity, came into being on March 10, 1853. On December 30, 1853, the charter for the NJH&D was sold to Pennsylvania Coal Company. When Pennsylvania Coal dropped all plans for NJH&D, the men originally involved repurchased NJH&D's charter from Pennsylvania Coal in January 1867. Back in the railroad business again, the NJH&D and three roads, New Jersey Western Rail Road Company, Sussex Valley Rail Road Company and Hoboken, Ridgefield & Patterson Rail Road Company, merged into a new corporation, New Jersey Midland Railway Company, in 1871. By May of 1871, the railroad was carrying freight and passengers. The closest this railway ever came to Pennsylvania coal was Franklin, New Jersey, still 15 miles east "as the crow flies" of the Delaware River which forms Pennsylvania's border.

New Jersey Midland became a part of the New York & Oswego Midland, which eventually went into bankruptcy. New Jersey Midland lasted until May 15, 1880, when it was sold to a new company, Midland Railroad Company of New Jersey, made up of New Jersey Midland's first mortgage bondholders. (Garret A. Hobart, a director of New Jersey Midland, had been appointed one of the two receivers for the railroad in 1875. Once the sale to the mortgage holders was completed, Hobart was designated president of Midland R.R. Co. of New Jersey. In 1896, Hobart was elected Vice-president of the United States as William McKinley's running mate.)

The new Midland road was still 15 miles from Pennsylvania and over 65 miles from any anthracite coal. In early 1881, rumors began to float that yet another new company had been formed which would be responsible for extending Midland rails from Ogdensburg, New Jersey, through Sparta and Newton to the Delaware Water Gap, and from there to the Scranton-Pittston area - anthracite coal at last. The name of the new company was the New York and Scranton Construction Company.

But suddenly a total of 28,000 acres of coal land was pledged to Midland of New Jersey, on condition that Midland go to Tamaqua, Pennsylvania. Tamaqua was in Schuylkill County. (Railroads made money carrying freight. For a railroad, the most prized of all contracts, because they represented so much freight revenue, were those with coal operators for carrying coal.)

On June 10, 1881, The New York, Susquehanna and Western Railroad Company (NYS&W R.R.) was formed. It was to be a continuous track from the Pennsylvania coal region to tidewater at Jersey City, New Jersey.

The NYS&W was made up of all the railroad companies necessary to go the distance: Midland; by lease, Middletown, Unionville & Water Gap R.R., Patterson Extension R.R., Midland Connecting R.R., North Jersey R.R., Water Gap R.R.; and the special company formed in Pennsylvania, Pennsylvania Midland Railway. Pennsylvania Midland's track was to begin at Stroudsburg and extend to an unspecified point in the coal fields. And just where was NYS&W going to tap into anthracite coal? Well, NYS&W was so busy putting all of its track in good shape to cross New Jersey that development of the Pennsylvania portion lagged behind.

That was because on August 6, 1881, New York & Scranton Construction Company (NY&SC) had completed the last of several agreements. By the August 6 agreement, William S. Dunn, president of NY&SC, had obtained the exclusive right to carry 200,000 tons or more of anthracite annually by 1886 from - you guessed it - William Winton and Edward Dolph and their mines in the Mid-Valley, at Jessup and Winton. A like contract had been executed with Pennsylvania Anthracite Coal Company - and you guessed it again - headed by Horatio S. Pierce as president.

For Winton and Dolph, the Construction Company would lend $150,000 for building two breakers and opening mines. Repayment of the loan would come in installments deducted from the sale of the coal. New York & Scranton was responsible for laying railroad track from a connecting point on its main line to the coal breakers, supplying a yard engine and sufficient coal cars for carrying away the coal. Winton and Dolph, in addition to mortgaging these particular coal lands, were committed to laying enough yard track/sidings to store the cars, with appropriate switches. As owners of the land, the two men would receive 25¢/ton. Whoever mined the coal would receive 50% of the market price at tidewater.

Then at the end of 1881, Frederick Potts, president of NYS&W, announced, "The cost of constructing a rival line in that territory would have been very great, as it is up over the Pocono mountains, and if the business could be brought from Scranton to Spragueville over the DL&W R.R., it would save a large amount of money for the NYS&W. Furthermore, it would add to the revenues of the DL&W R.R., and afford the people of Scranton the advantage of two routes and outlets to the seaboard."

(Spragueville, at one time known as Analomink, where William Henry had established iron works about 1830, was in Stroud Township, Monroe County, just north of Stroudsburg. The changeover between the DL&W and the NYS&W track actually occurred in an area that came to be known as Gravel Place, where Brodhead Creek forms part of the boundary line between Stroud Township and East Stroudsburg Borough.)

Ah, ha! Tamaqua was forgotten. NYS&W cars would be carrying coal from Scranton and Pittston as originally planned. But again, like Lackawanna County and the Second National Bank - at a price. The New York & Scranton Construction Co. reported, "This was a good deal of surrender for the New York & Scranton Construction Company, but it provides a temporary relief from the financial difficulties surrounding them." It had been agreed that with the DL&W entering the picture, the New York & Scranton was no longer necessary.

Or was it?

On January 7, 1882, Dunn, for NY&SC, approved a change in the earlier agreements. Old friendships, going back to the mining days of Wintons' New York and Pennsylvania Coal Co. in the 1850s, had been invoked; John Jermyn was now superintendent of New York & Scranton Construction Company. Jermyn would supervise operations at Winton and Jessup for the NY&SC. Jermyn, by the way, was general manager of Pennsylvania Anthracite Coal Co.

Then the great day finally arrived, on October 26, 1882; the new rail line opened. The DL&W brought the coal-filled cars to Gravel Place. There, they were hooked onto NYS&W engines and pulled to West End, Jersey City, where DL&W equipment completed the journey to its coal docks at Hoboken. "The combination" had no authority over this arrangement, and it showed in coal tonnage totals. 1883 was the last year in which more coal was shipped from the combined Schuylkill and Lehigh regions than from the Wyoming region (Lackawanna and Luzerne Counties). Starting in 1884, the Wyoming region alone shipped more anthracite than the other two areas.

How was it done?

Chapter 52

The legacy that was handed over to their youthful replacements called for them to be leaders in economic activities.

When the Second National Bank closed its doors, other banking institutions in which William Winton had an interest also found it expedient to shut down.

Feelings among some depositors, who feared they would not be repaid, ran so high, especially in what had once been the Providence Borough section of Scranton, that a bust of William Winton placed at the drinking fountain in Providence Square, in front of Winton's Citizens and Miners' Savings Bank and Trust Company of Providence, was knocked over and torn from its pedestal. (Newspaper headlines screamed, "Bust of Busted Bank President Busted!" The statue was apparently not totally destroyed, because 20 years later it was found stored in Winton's barn/carriage house.) In another act expressing dissatisfaction, Winton was hanged and burned in effigy. Feelings were running very high and intense. There was also a deep affliction running through the members of the family who had supported Winton, especially his brother-in-law Isaac Dean. It did not seem to matter that all the Winton financial institutions that closed, including the Second National, had done so at a time when their assets were greater than their liabilities; that no depositor would be wiped out. The intense emotions of the moment seemed only to serve the purpose of exacerbating the situation, as if asking for panic, and financial ruin for individual depositors.

Neither William Winton nor any of the men closely associated with him had expected, or believed, that the Second National could go under. "Business as usual" was the watchword, as the days of 1878 progressed to that fateful June 1 and the shutting of the bank's doors. Unfortunately, what made it all the more devastating was that the closing of the Second National came at a very critical moment - when plans for future activities were about to go forward.

So now a period of concealment and subterfuge by the Wintons and those around them began. The handling of the contracts with the New York & Scranton Construction Co. (NY&SC) was one of the first examples. Of course, there had been a second purpose in maintaining a low profile in regard to NY&SC - to keep as much information as possible from reaching members of "the combination."

In the past, Elisha Phinney had often served in high visibility roles. Phinney had been in on the Pennsylvania Anthracite Coal Company, Lackawanna & Susquehanna Coal & Iron Co., Stowers Pork Packing & Provision Co., and had worked on establishing a final resting place for all the old friends together: Forest Hill Cemetery, on the Scranton-Dunmore border. As recently as February 18, 1878, Phinney had traveled to Somerville, New Jersey and there formed a gold and silver company that would operate in Georgia, north of Atlanta. Phinney was an integral part of all the activities that focused on Scranton's and northeastern

Pennsylvania's role as the heartbeat of America's Industrial Revolution.

With Phinney out of town, with Winton almost a pariah, and Horatio Pierce and John S. Law forced into maintaining very low profiles, the task of moving forward fell to competent, sometimes youthful replacements. Law, who was probably the least vulnerable, shouldered most of the burden for keeping things going.

The three men, Pierce, Law and Winton, put coal mining activities on the 180 Blakely acres they had purchased in 1875 into a newly formed company, The Lackawanna Coal Company, Ltd., in June 1881. Lackawanna Coal and its officers then became the visible part of the activity that would mine coal and send it to tidewater in New York, Susquehanna & Western (NYS&W) railway cars. By the following year, John S. Law was running Lackawanna Coal as president. Edward B. Sturges was secretary (Sturges was George Sanderson's son-in-law). Orlando S. Johnson was the company's superintendent. Sturges and Edward N. Willard served on the Board of Directors with Pierce appearing as the third member of the board (no Winton). Lackawanna Coal was to be responsible for producing at least 50,000 tons of coal a year for every year after 1883.

In September 1882, Catherine Heermans Winton, joined by her husband, William Winton, leased a 20 acre tract in Scranton City for mining to Orlando S. Johnson, Edward N. Willard and Louis A. Watres. (Watres was the son of Lewis S. Watres who had settled in the 1830s at Mt. Vernon (later Winton). L.A. Watres had just been elected to the Pennsylvania Senate for the first time in the 1882 elections.) This coal would also go to tidewater in NYS&W coal cars. Almost 16 years later, Johnson would still be mining the Scranton City property under the name of The Green Ridge Coal Company, with himself as president and William Lawrence Connell as secretary. (William L. Connell is not a direct descendent of William Connell.)

In the meantime, the moving force behind all of Sullivan County, Michael Meylert, would die in January 1883. But Meylert and his family would continue to be a part of the long years of association and its legacy, even though he and his wife had been childless. His niece Ada, daughter of his brother Amos N. Meylert, had married Joseph Augustine Scranton in 1863. This Scranton, a son of the Joseph H. Scranton who had come up from Georgia to invest in the iron works in the 1840s, and a step-brother to William W. Scranton, had started the Scranton newspaper *Daily Republican*. And he had thrown his newspaper resources behind the fight for the new county, publishing the issues calling for the new county that were placed on the desks of legislators at Harrisburg.

And John Heermans, who had taken his family in 1865 to Corning, New York, while serving as private secretary to his uncle Joseph Fellows, Jr., was nearing the end of his own life. In February 1881, Heermans turned over administration of his uncle's estate to William H. Jessup, the son of William Jessup. Jessup would serve in the 1880s as "counsel for the Erie Railway, Delaware, Lackawanna and Western Railroad, Delaware and Hudson Canal Company and Montrose Railway, besides holding several important trusts, the most extensive of which is the trusteeship of the Joseph Fellows estate."

The legacy that was handed over to their youthful replacements called for them to be leaders in economic activities.

This meant accumulating money, and investing and managing it wisely. Those who showed the ability to be leaders were to be respected for their leadership. But at the same time, there would always be an unwillingness for anyone to show or admit to being impressed by a great personality. Anticipations of economic advantage would always be sufficient motive for undertaking commercial ventures, but only to the extent that the personal gain was not a rape. Money was to be used and used wisely. Money did not exist for lavish displays. In this philosophy, there was little need for public contributions or for welfare. The individual was a totally responsible person who looked after his neighbor, the family after its members.

(While religion and religious affiliation were not a consideration, nor ever mentioned, this was a legacy that was instinctive with Quakers.)

If this was their legacy, how would the youthful replacements and the next generation handle their responsibilities? Would Pierce, Law and Winton be able to continue their economic leadership?

Winton continued to exercise some leadership because he would not be stopped. He formed Winton Coal Company, Limited, in 1883. He, his wife, two of his four sons, A.H. and Byron, and his son-in-law Thomas Livey were the shareholders in the company. But it was his wife, Catherine Heermans Winton, who held three-quarters of the stock. The injunction which prevented a deed from the Sheriff of the new Lackawanna County conveying the properties Catherine Winton had bought in her name from being recorded was still in effect. To get around this dilemma, in 1885 both Catherine and her husband deeded all real estate listed in the sale into Winton Coal Company. Edward Dolph also formed The Dolph Coal Company, Limited, for his properties at Winton. Two of his fellow shareholders were Edward B. Sturges and Isaac Hand (who had married a grandniece of Joseph Fellows, Mary Heermans Richardson. Hand, himself, was a great-nephew of Alfred Hand).

The pattern was set; relatives and friends, and the sons of relatives and friends, would be the ones to move forward as leaders in economic activities. Isaac Dean's son, Arthur Dean, had become a lawyer. Arthur was Catherine Heermans Winton's nephew. Edward Laton Fuller and Winton's son, Byron Winton, were brothers-in-law, both having married daughters of Daniel Silkman. John S. Law's nephews, and the husbands of his nieces, would make the greatest contribution numerically: Archibald F. Law, George W. Cross, and Thomas H. Watkins with his business partner Clarence D. Simpson. Elisha Phinney's son-in-law, Charles Pooley Matthews, came forward, too. These were all men willing to assume the legacy passed to them. Others, such as James Everhart, Willoughby W. Watson, Mahlon S. Kemmerer, Thomas

Waddell, Thomas J. Foster, J. Wilson Peck, and William Connell would join in. Connell, who had been active in coal mining from the 1850s and a partner in the Hunt brothers mercantile operation (which changed to Hunt & Connell, Limited by 1886) became even more involved on the death of Horatio S. Pierce in 1889. Pierce had only a daughter to inherit. He placed his interest in Lackawanna Coal Company in the hands of James S. McAnulty, until his grandson came of age. McAnulty was a son-in-law of William Connell.

Where would these men take the legacy that had been passed to them?

Chapter 53

Now, more than ever, loaning start-up capital, bonds and/or capital stock became the vehicle by which they remained active.

On the coal lands they owned, Horatio S. Pierce, John S. Law and William W. Winton opened as many veins of anthracite coal as they possibly could. A multitude of coal companies sprang up to work the veins; almost 30 mines were opened. By the end of the decade, in December 1889, over three million tons of coal had been added annually to the northern coal field export figure, making a total of 8,249,087 tons mined. This represented the work of 6,240 men, a quarter of the total mining work force of 23,625 in Lackawanna County. But few knew that Pierce, or Law, or Winton were involved.

Now, more than ever, loaning start-up capital, bonds and/or capital stock became the vehicle by which they remained active.

Winton and Edward Dolph had sold coal to the New York & Scranton Construction Co./New York, Susquehanna & Western

R.R. before mining it, to borrow from the construction/railroad company, in order to erect breakers and set up mining operations. But to have the money to lend, the railroad had raised money by selling bonds - and who do you think were some of the bond buyers? Like the proverbial puzzler - which came first, the chicken or the egg? - the situation begs asking - who was borrowing money from whom?

For Winton, when mining operations actually started in 1886, the shaft was known as the S.V. White Tunnel. Stephen V. White, from Brooklyn, New York, had replaced William S. Dunn as vice-president of New York, Susquehanna & Western R.R. (NYS&W). (Dunn had been president of New York & Scranton Construction Co. prior to moving over to NYS&W.) It was White who, on November 2, 1885, agreed to supply the NYS&W, by lease arrangement, with 200 20-ton drop bottom coal cars and two Mogul locomotives, #36 and #37, made by Rogers Locomotive and Machine Works of Patterson, New Jersey, "making the aggregate value of all equipment hereby leased, $89,300." But it would be Winton's son, A.H. Winton, who would record in his seven-year old daughter's diaries, for Friday, October 9, 1885, and on the page for Monday, January 4, 1886, the cryptic notations of a meeting between Frederick A. Potts, Stephen V. White, Horatio S. Pierce, John Jermyn and the Wintons (prior to White leasing the locomotives and coal cars to NYS&W), and three months later that the first shipment of eight cars had left the S.V. White Breaker.

Much of the mining operations from 1886 on, however, were entrusted to members of the Law family. Thomas H. Watkins, married to Elizabeth Law (John S. Law's niece), with partner, Clarence D. Simpson, opened and operated over six collieries, including Grassy Island, Babylon, Forty Fort, and Northwest Coal Company. Watkins' brother-in-law (who was also John F. Law's nephew) Archibald F. Law, coming from a stint as cashier of the Canada Southern Railway at Buffalo, New York, would join him and Simpson.

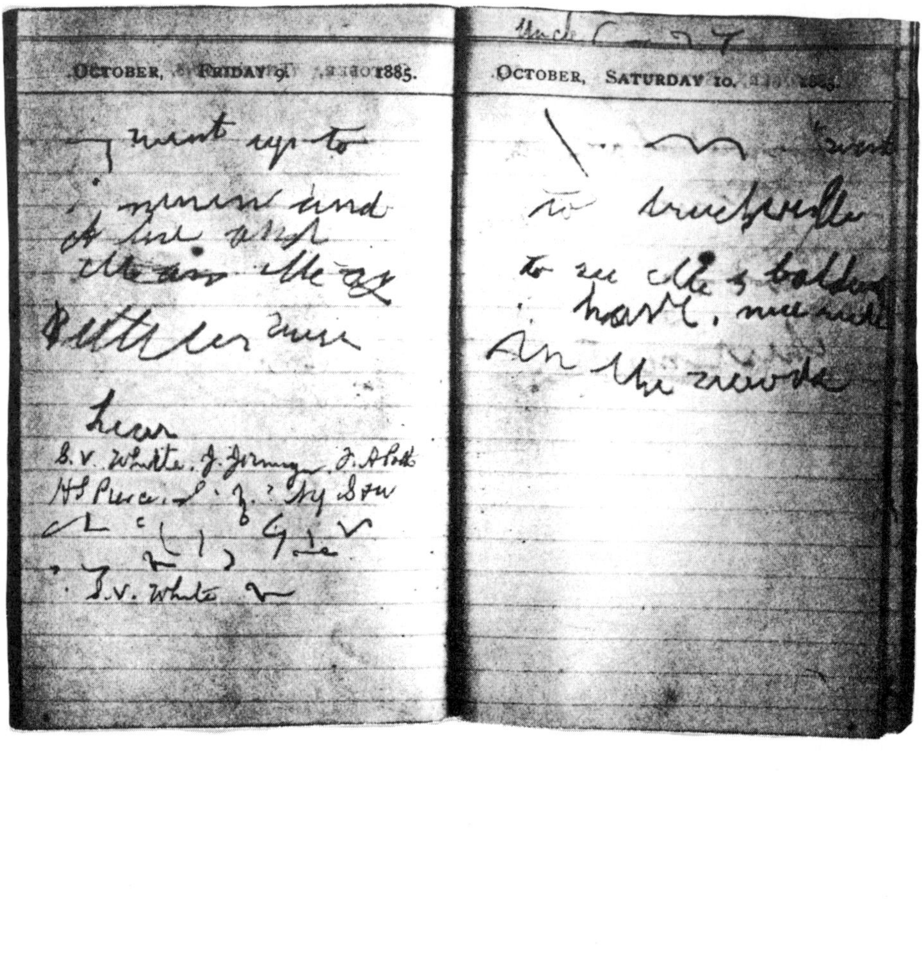
OCTOBER, FRIDAY 9. 1885.
OCTOBER, SATURDAY 10.

DECEMBER, 1886.

Time explained on page preceding Jan. Calendar.

FIRST QUAR. D. H. M.	FULL MOON. D. H. M.	LAST QUAR. D. H. M.	NEW MOON. D. H. M.
3 9 26 M.	11 4 31 M.	18 1 40 M.	25 4 56 M.

DAY OF YR.	DAY OF MO.	DAY OF WEEK	SUN RISES H. M.	SUN SETS. H. M.	MOON R. & S. H. M.	HIGH TIDE. BOSTON H. M.	HIGH TIDE. N. YORK H. M.
335	1	Wed.	7 5	4 33	10 20	2 57	*morn*
336	2	Thu.	7 6	4 33	11 16	3 43	0 41
337	3	Fri.	7 7	4 33	*morn*	4 30	1 29
338	4	Sat.	7 8	4 33	0 2	5 20	2 18
339	5	SUN.	7 9	4 33	1 8	6 12	3 9
340	6	Mon.	7 10	4 33	2 5	7 4	4 1
341	7	Tue.	7 11	4 33	3 3	7 56	4 52
342	8	Wed.	7 11	4 33	4 0	8 46	5 43
343	9	Thu.	7 12	4 33	5 2	9 35	6 33
344	10	Fri.	7 13	4 33	6 2	10 23	7 20
345	11	Sat.	7 14	4 33	*rises*	11 10	8 5
346	12	SUN.	7 15	4 33	6 15	11 56	8 54
347	13	Mon.	7 16	4 33	7 18	*eve* 55	9 43
348	14	Tue.	7 16	4 33	8 24	1 37	10 33
349	15	Wed.	7 17	4 34	9 31	2 28	11 22
350	16	Thu.	7 18	4 34	10 39	3 19	*eve* 17
351	17	Fri.	7 19	4 35	11 48	4 14	1 13
352	18	Sat.	7 19	4 35	*morn*	5 14	2 13
353	19	SUN.	7 20	4 35	0 53	6 13	3 11
354	20	Mon.	7 21	4 36	2 0	7 17	4 14
355	21	Tue.	7 21	4 36	3 7	8 15	5 12
356	22	Wed.	7 22	4 36	4 12	9 10	6 8
357	23	Thu.	7 22	4 37	5 16	10 3	7 1
358	24	Fri.	7 22	4 38	6 17	10 51	7 45
359	25	Sat.	7 23	4 39	*sets*	11 37	8 32
360	26	SUN.	7 23	4 39	6 13	*morn*	9 19
361	27	Mon.	7 23	4 40	7 10	0 20	10 0
362	28	Tue.	7 23	4 41	8 7	1 2	10 39
363	29	Wed.	7 24	4 41	9 3	1 45	11 19
364	30	Thu.	7 24	4 42	10 1	2 25	*morn*
365	31	Fri.	7 25	4 42	10 57	3 6	0 3

1st Shipment of 8 Cars

JANUARY, 1886.

FRI. 1. Ther. Weather,

DAILY MEMORANDA.

Space is provided for recording the state of the Thermometer and weather. The Moon's Changes, and Stated Church Days are also noticed as they occur.

Signs: New Moon; First Quarter; Full Moon; Last Quarter.

Uncle walter came down from providence and took dinner with [illegible] and Fannie bell mortimer [illegible] miller and [illegible] and I and uncle [illegible] all went [illegible]

JANUARY, 1886.

MON. 4. Ther. Weather,

in the morning Adee [illegible] and helen Matthews came over to see me and in the afternoon I went over to see them

1st Shipment 8 cars to S.V. White Brakes

JANUARY, 1886.

TUES. 5. Ther. Weather,

helen and Adee Matthews came over to my house in the morning and in the afternoon. in the afternoon

14 [illegible] 10 [illegible] completed

Elisha Phinney's son-in-law, Charles P. Matthews, continued to play a role as Phinney had. And Matthews' son, Willard Matthews, would marry Cora Brooks, a daughter of the well known coal operator Reese G. Brooks. For his part, Charles Matthews brought in former Susquehanna resident Willoughby W. Watson, E. B. Leisenring, and Mahlon S. Kemmerer to form Mount Jessup Coal Company, Ltd., and Moosic Mountain Coal, with spinoffs Moosic Mountain and Carbondale RailRoad and Moosic Mountain Water Company, all for supplying and mining on Moosic Mountain in the Winton and Jessup to Olyphant areas.

Leisenring and Kemmerer were Carbon County/Mauch Chunk/Jim Thorpe names. Leisenring, before becoming active with Moosic Mountain Coal Co., had served as mine superintendent for Lehigh and Wyoming Coal Company at Audenried, Banks Township, Carbon County, in the 1870s. Kemmerer, starting Mount Jessup Coal Co. November 17, 1888, had married another Leisenring, Annie, the daughter of John Leisenring. John Leisenring had been a protegé of Josiah White; he later lived up to White's expectations when he became president of Lehigh Coal & Navigation Company. Kemmerer and his father-in-law had jointly taken an active part in the Carbon Iron Company, starting in 1877. Another venture of Kemmerer's, after starting Mount Jessup Coal Company, was to travel to southwest Wyoming, where he became co-founder in 1889 of the mining town of Kemmerer, seat of Lincoln County. (The world's largest open pit coal mine is at Kemmerer.)

Another incorporator of Mount Jessup Coal Co. was James Everhart. Everhart was interested in mining generally. (His brother, Dr. Isaiah F. Everhart, established the trust for the Everhart Museum in Scranton.)

In the 1880s, there were, however, other things to consider. Electricity was fast becoming an important buzzword. Corning Glass was busy turning out the glass casings for Edison's light bulbs in 1881. And if you think the boys in Scranton were oblivious to what was going on at Corning, New York, just because Joseph

Fellows was no longer there, forget it. John Heerman's son, Harry C. Heermans, was as close to his Scranton cousins as Fellows had been when he first started to work for Troup and Pulteney Associates. So, yes, electricity and its applications were of special interest.

Thomas Alva Edison completed installation of an electric power station in New York City in 1882. Electric companies did not show up in Scranton until 1884 - Scranton Electric Light and Heat Company - with William W. Scranton, son of Joseph H. Scranton, as president of the company. As Edward Sturges would write later, "For years I had believed that electricity was the coming power." It was under Sturges' actions that Scranton finally forged ahead in 1886 and took the lead in electric transportation, with an electrified trolley car line. The innovation earned the title of "The Electric City" for Scranton.

How did it come about?

It was certainly one of the world's best kept secrets.

A reliable chronicler of Scranton's past, Frederick L. Hitchcock, who had lived all his life in Scranton, and as a young man served in the Union Army in The War Between the States/Civil War, was sufficiently foggy on the matter himself, and therefore asked Sturges in 1913 for history's account.

Keep in mind when reading Sturges' letter that the publicly announced purpose of Scranton Suburban Railway Company, incorporated at the beginning of 1886, was "to build a rapid transit railway to Green Ridge and Forest Hill Cemetery." And then remember that almost half of Forest Hill Cemetery would still be empty for another five or more years, as Wintons, Dolphs, Sandersons, Fullers, Fosters, etc. were actively running around the streets of Scranton.

"July 9, 1913

"My Dear Colonel [Hitchcock]:

"Your questions as to the Scranton Suburban Railway reached me after I had returned from Connecticut. Have been waiting until I could get more time before replying to them. Even now I am not able to write you as fully as I would like. I will, however, reply in a general way to what you seem to have in mind, which as I understand it is to ascertain when Scranton's first electric railway was built. The Scranton Suburban Electric Railway, of which I was President and the principal stockholder, actually commenced operation in November, 1886, and I claim it to have been the first successful electric railway operated in the United States.

"There had been in several places attempts to operate street car lines by electricity, but at the time when ours was started every one of them - with one possible exception - (that at Appleton, Wisconsin) - had been abandoned. Mr. O.S. Johnson [Orlando S. Johnson] and myself, about the time we commenced the construction of the Scranton Suburban (approximately April, 1886) went to every point in the United States where electricity had been tried. Among others, I remember Montgomery, Alabama, Richmond, Virginia, South Bend, Indiana, but in none of these places was there any attempt to operate by electricity, all such attempts having resulted in failure.

"Port Huron is not in the United States but in Canada, and along the Detroit River, nearly opposite that city. My recollection is that no cars were operated even there, but if so it was in a very small way. In every other place there had been an entire abandonment of the attempts.

"After we had begun the actual building of our road (which as you remember stopped behind my house), I went to Europe for the purpose of learning whether any electric roads were being successfully operated there. Among other places I went to the extreme north of Ireland, to the Giant's Causeway. The road there was not in operation, apparently abandoned. I went to Frankfurt, Germany, at which point an occasional car was said to be run over a short road near that city, but I did not see any.

"Bude-Pesth [Budapest], the capital of Hungary, had an extremely advanced underground electric system, something similar but of course very imperfect - to the present New York underground conduits, but that was not being operated. They were rebuilding it when I was there, hoping to remove some defects, so that as far as I can remember I never had a mile travel anywhere upon any electric road until our own Scranton road was started. It is true, however, that when we ordered our cars from the Pullman Co. (the finest cars ever built for street service, even up to this date), the Pullman people were building a couple of cars for Appleton, Wisconsin (a lumber district). This road started about the same time as ours, but whether immediately successful or not I have never been able to learn. It was not an expensive system, but one running out to a suburb.

"In several magazine articles that I have seen, Richmond, Virginia was given the credit for having the first successful railway in the United states, but careful investigation has satisfied me that Richmond was at least six months behind us in any successful operation by electricity. I have timetables from our suburban road antedating [by] more than six months any claim that Richmond has ever made that I have heard of.

"I am perfectly well aware that my life has not been in all respects that of a pioneer, but I do claim to have inaugurated and largely paid for the most successful electric railway in the United States, out of which I did not make one cent in money. Not being a street railway man or a financier, I sold it for exact cost plus six percent interest to a Mr. Bacon of New York City, who was acting for parties the names of whom I cannot now remember. It soon, however, got into the possession of some Philadelphia people who had gotten control of the Peoples Street Railway lines and electrified them.

"There is no doubt whatever in my mind that Scranton is entitled to the credit for having inaugurated the first electric railway, and I claim the credit for having inaugurated the first electric railway, and I claim the credit for having been the principal man who secured it for the city. It was Mr. Charles J. Vanderpoele of Chicago, whose system was used by us, and to him I have always claimed belongs the chief credit for the present magnificent electric railway system of this country. The Thomson-Houston people who afterwards organized the General Electric Company, were shrewd enough to see the value of Mr. Vanderpoele's inventions and secured a permanent contract for his employment, which continued up to the time of his death - a few years after the system had been fully developed. For years I had believed that electricity was the coming power, and I either furnished the funds or secured subscriptions to the funds that built the Scranton road.

"Let me know if I can be of any further service to you, and with regards to you and yours, I am,

"Very very sincerely,

"Edw. B. Sturges"

(George Sanderson, Alfred Hand, Merle J. Wightman, and Sturges were among the men who formed The Wightman Electric Manufacturing Co. in 1890. Later, the controlling interest of the company was purchased by the General Electric Co. Wightman was generally recognized as an electrical inventor. In 1893, Sturges was serving as a trustee for the Comstock mining operations in Nevada.)

Many thanks to Attorney Edward Sturges and his letter for posterity. But still the beginnings of "The Electric City" remain shrouded. Scranton Suburban Railway was capitalized at $20,000, yet Sturges and Johnson incur a proportionately large travel expense just for traveling all over the United States and Sturges (alone?) across the capitals of Europe - for what? To run a rapid trolley line to a cemetery? Which will be using the finest Pullman Company cars ever built for street service, with interior electric lighting? It should be kept in mind that before Sturges left on his travels in April, he and Catherine Heermans Winton jointly took title to a one-third portion of 165 acres of very rich coal land in Jessup, on March 2, 1886, for $1,000. The property was sold to the new co-owners by the receiver for the Second National Bank. And on October 22, 1886, just one week before the new trolley's trial run, Edward Sturges, along with William W. Winton and one of his sons, William H. Winton, appeared in the Recorder of Deeds office for Lackawanna County to record the incorporation of a new company, The Grassy Island Water Company. In the notarization, all three men signed their names to a statement that they had appeared together before the Recorder of Deeds on October 22, 1886.

A test run was made on the new trolley line on November 29 and commercial service began on November 30, passengers apparently preferring to avail themselves of the opportunity to hear Henry M. Stanley recount his adventures in Africa seeking Dr. David Livingston, rather than exercise their rights under the avowed purpose of Scranton Suburban Railway and ride to Forest Hill.

Electrically-propelled passenger transportation was not the only new field where advances and explorations were taking place. Some, like a new company incorporated in West Virginia, would become an integral part of industries that would eventually contribute to a downsizing of anthracite mining. In 1883, stockholders of the Barber Asphalt Paving Company, namely J.J. Albright, James Archbald, Joseph J. Albright (names sound familiar?), E.B. Warren, and A.L. Barber, called for election of officers and directors at the company's main office in Washington, D.C. Yet a branch office of the asphalt paving company was not opened in Scranton until 1886. By 1891, the central part of Scranton City's streets had been paved, a total of 103,002 square yards.

In still another direction, Louis A. Watres, usually with Lemuel Amerman, starting in 1886 incorporated more than 12 different community water companies. Four of the companies were registered on the day before Christmas: December 24, 1888. Of the four, Winton Water Co. soon had William W. Winton as president. Watres moved onward and upward; in the 1890 general elections, he was elected lieutenant-governor of the Commonwealth of Pennsylvania (and was a very popular public official).

Even as Lieutenant-Governor, Watres was still out there forming water companies. In September 1891, he and Amerman were in Tioga County - you guessed it - The Blossburg Water Company was founded, followed by The Mansfield Water Company in 1893.

Watres' sudden craze for forming water companies was not so strange, nor an isolated aberration, when you consider Harry C. Heermans at Corning, New York. This particular Heermans had often assisted his father, John Heermans, in his duties for John's uncle Joseph Fellows. After graduating from Wesleyan University in 1875, Harry returned home to Corning to study law. Soon, the concerns of the Corning water system reached Harry Heermans' ears. The system had been built in 1871 for $25,000, and by 1876 had had another $40,000 put into it, but was still "a source of continual vexation and expense to the village [of Corning]." No one would take over the system, and there the matter stood until Harry Heermans, "with [partner] Thomas Lawrence, proposed to lease the plant for a term of years and assume all responsibility for its proper management, extension and maintenance, though he met with discouraging advice from his best friends and the strongest possible objection from his father [John Heermans]." The "village fathers" succeeded in getting Harry up to a 30 year lease starting on January 1, 1877. "Thereupon, by the investment of private funds and in the face of many obstacles and discouraging objections the works were placed on a successful basis." It took the two partners three years to bring the water company to where it was breaking even. After that, the water works became profitable and was soon recognized as "one of the best enterprises in Steuben County," pumping almost one million gallons of water per day. From then on, it was one success after another for this Harry Heermans. "In 1886 he devised a plan and constructed the water works system of Wellsboro, Pennsylvania, and in 1893 a similar works in Westfield, Pennsylvania," both in Tioga County (Wellsboro is its county seat).

Harry Heermans and his cousin John H. Fellows also went off to the northwestern United States. There, they would be busy off and on with a variety of projects, while their relatives expanded into the southern and southwestern United States and elsewhere.

Chapter 54

So as the 20th Century drew nearer, more attention was paid to manufacturing, and business activities other than anthracite mining.

Banking, however, remained a problem. For its resolution, new banks were started; also funds were scattered among new and long established banks.

Horatio S. Pierce fulfilled his duties as assigneé of The Scranton Trust Company & Savings Bank for about a year; he in turn assigned winding up of the bank's affairs to James A. Linen and George Fuller in March of 1880. The only difference between Pierce and the other two men was that Pierce had taken over management of all The Scranton Trust's assets, both real and personal (land and stocks, bonds and mortgages/loans), whereas Linen and Fuller took over only "all the real estate of the said The Scranton Trust Company and Savings Bank." Linen had become cashier of the First National Bank of Scranton shortly after the end of The War in 1865. He held that post until 1891, when he became president of the First National. Linen was a first cousin of Thomas Dickson, who started Dickson Manufacturing Company and served as president of the D&H Canal and Rail Road. Linen had married Anna Blair, a daughter of James Blair, thereby becoming a nephew to the "Railroad King of the West," John I. Blair. Linen had also been a part of the Scranton Stove & Manufacturing Co. under Pierce, the Hunt brothers, and William Winton. Arthur Fuller, a distant cousin of the Edward Fuller family, was still active with Scranton Stove, having become a co-owner, when named an assigneé of Scranton Trust Co.

The long-established First National became one of the banks to fill in where Winton's banks had once held sway alone. Another older bank, The Scranton Savings Bank and Trust Company (distict from the Scranton Trust Co. & Savings Bank) had started at the beginning of 1874, with Edward N. Willard as president and Louis Watres as cashier. The two men continued their association with the bank for years, although Watres moved into the secretary's slot. Eventually, Orlando S. Johnson joined the board of directors. Then there was the Third National Bank, organized in 1872. Alfred Hand, William Jessup's son-in-law, and later William Connell, both served there as president. When the Third National first opened, William Winton was listed as a director. *Today,* some of Winton's descendants continue to hold shares of Third National's stock.

A new bank, organized in 1887, was the Lackawanna Trust and Safe Deposit Company. Its directors included Edward B. Sturges, George Sanderson, William Connell, Charles duPont Breck, and Henry Belin. Another new bank, the Traders National Bank of Scranton, first opened its doors on the first business day of 1890, using the Second National Bank building. This bank had J.J. Jermyn, James Everhart, Charles P. Matthews, Willoughby W. Watson and Mahlon S. Kemmerer and others, on its board of directors. The president of Traders' was Samuel Hines, who had been associated with the Blossburg Coal Company. So, there were many banks in Scranton ready to receive deposits. Outside of

Scranton, Quincy W. Wellington, at Corning, New York, was fast getting a reputation "as one of the most successful business men and bankers of the region." In 1884, Wellington made his son Benjamin Wellington a partner in the banking firm of Q.W. Wellington & Co. On January 4, 1886, William H. Jessup transferred his trusteeship of Joseph Fellows' estate to Quincy W. Wellington.

It didn't seem to matter too much to these men that they no longer had "their own bank." William W. Winton was still living, but Horatio S. Pierce was gone, and John S. Law had left the area to live in New York City. So financial matters were divided up among the men who were already united in thought and deed (and also by blood or marriage). Each man safeguarded the interests of all, at whatever banking institution he was most familiar with.

It was in the 1880s, as money was scattered among several banks, that the public perception of Scranton and northeastern Pennsylvania became that of an anthracite coal region, a mining area, and nothing more; that other than the Lackawanna Iron & Coal Co. and the homegrown DL&W and D&H railroads, there was no financial power base in that section of the Commonwealth. But the public image couldn't have been further from the truth. The men who had come along where Pierce, Law and Winton had once been seldom allowed any one company or business interest to accumulate and grow into a giant of a corporation. Where there could have been one large mining company, there would instead be ten mining companies, the men running the ten the same men who operated the one original company.

This business practice was the opposite of what was taking place in the bituminous (soft) coal region, centered at Pittsburgh's iron and steel mills. There, Andrew Carnegie, Henry C. Frick and later Charles Schwab, were dedicated to a course of action that would create the largest single business possible. They finally achieved a form of success when negotiations in 1901 led to formation of a holding company, and the United States Steel Corporation began operations in 1902. (Carnegie was bought out by New York financier J. P. Morgan and his partners for $400 million.)

Another factor that the men in northeastern Pennsylvania who dealt in natural resources were fully aware of was that their products were on a downward spiral; someday these natural resources would be totally depleted. Estimates were often published that there were enough tons of anthracite for another 200 years, or whatever combination of numbers the writer had accessed. But as the new 20th century approached, more interest was paid to other possible activities. William W. Scranton, when he proposed a toast to "Our Manufacturing Interests" at the banquet celebrating the erection of Lackawanna County and the laying of the Courthouse cornerstone on May 25, 1882, sounded the warning bell:

"Now, I don't hold myself as a great many do about the town, but it seems to me that this community is based principally upon the mining of coal - half of it depends upon it. Now these mines are being worked out more and more, and, in my mind, it is not going to be forever. It will not be one hundred years before these are all worked out. To me it seems that a great many of us here are going to live to see the time when the growth of this town, so far as it depends upon mining, is not going to increase. When that time comes, we have got to fall back upon something else. It seems to me that we ought to provide for that beforehand, and provide, as far as possible, for our manufacturing interests. For that, we want good government; we want law and order; we want modern taxation."

So as the 20th Century drew nearer, more attention was paid to manufacturing, and business activities other than anthracite mining.

Several companies that had been formed in the 1870s now surged forward. The Sauquoit Silk Manufacturing Company led the way. Sauquoit had started as a manufacturer of silk thread only. In 1885, the company added the making of silk cloth to its line of goods. Two other silk companies started up: Harvey's Silk Mill and Meadow Brook Silk Co., Ltd. Once the advantages of employing the women in miners' families were recognized, knitting mills became popular. In 1887, William Connell appeared as president of the new The Lackawanna Mills (a knitting mill) and of another factory - Scranton Button Company, Limited. Three years later, Connell was offering land suitable for a factory site to the lace manufacturers from Nottingham - England, Mosely, Wotten & Clifton. Company representatives were given the "red carpet" treatment by Scranton's Board of Trade (now the Scranton Chamber of Commerce) in 1890. Interestingly, after Connell had to drop out of the British deal because the land he was offering was determined to be too small for the project, George Sanderson showed up the following year (1891) as treasurer for the newly formed The Scranton Lace Curtain Manufacturing Company.

From another direction, support was still coming from Joseph Augustine Scranton and his Meylert wife. Scranton was elected to the U.S. House of Representatives in 1880 (47th United States Congress). While in Congress, Scranton secured passage of the bill authorizing a federal building - post office - at the southwest corner of North Washington Avenue and Linden Streets. (*Today*, the federal building, rebuilt, is still there.) Scranton was re-elected to the 49th, 51, 53, and 54th sessions of Congress, carrying him to 1896. It was while a member of the 49th Congress that Scranton was able to bring United States District Court sessions to Scranton. (After Lackawanna County, having a U.S. District Court based in Scranton was probably A.H. Winton's greatest desire.)

One of the most important aspects of Scranton's service at the nation's capitol occurred when his election years coincided with those of the 23rd President of the United States, Benjamin Harrison - 1889-1893. Harrison had married Caroline Scott, sister-in-law to Russell F. Lord, of Honesdale. Russell Lord had been with the D&H from 1826, rising to Chief Engineer before his retirement in 1864. The Lords' daughter, Mary, had been married briefly to Walter Dimmick when he died in 1881. When Caroline Scott Harrison became ill while First Lady, Mary Lord Dimmick took up residence at the White House and acted as hostess in her aunt's place, continuing in that role after the First Lady's death in

1892. At the time northeastern Pennsylvania newspapers often mentioned area residents traveling to Washington for dinner, etc. at the White House. Mary eventually married Harrison, but in 1896 after he had left office.

The Harrison administration is best known for the changes it accomplished in the Monroe Doctrine. The Monroe Doctrine, in place from the days of James Monroe's presidency in 1823, was America's foreign policy of isolationism. Harrison reformulated the Doctrine, holding the first Pan-American Conference which established diplomatic and commercial ties with independent republics in Central and South America.

In February, March, and April 1891, A.H. Winton traveled to Mexico and California. The fruits of his trip would not ripen until after the turn of the century, in 1901 and 1902.

(John S. Law's nephew-by-marriage, Thomas Watkins, and Eli E. Hendrick, the old friend from the days of oil exploration in western Pennsylvania, had gone to Los Angeles, California earlier taking with them patents for ice-making machines.)

While Joseph A. Scranton was expanding the horizons for northeastern Pennsylvanians, an idea that had been translated into reality would soon make Scranton as well known internationally as its exports of anthracite coal. The idea of learning by mail was possibly the greatest contribution to marketing.

The newspaper *Shenandoah Herald* had been founded in 1870 by brothers-in-law Thomas J. Foster and Henry C. Boyer. The *Herald* served residents of the Shenandoah mining community in Schuylkill County, near the eastern end of the Western Middle Anthracite coal field. The *Shenandoah Herald*, by its daily edition *Evening World*, was "the paper that made the original and valiant fight against the Molly Maguires." The "Molly Maguires," or Mollies, was a name given to a secretive group of men that became prominent in Schuylkill County, starting in 1862. The Mollies were described as "inciting resistance to the officers of the law and raising riots to stop work at some of the collieries where miners had struck." After a period of inactivity, the general perception in

Schuylkill County was that the Mollies, starting "in December of that year [1871] commenced a fearful career of crime, that made their names infamous throughout the whole civilized world, by a murder in Carbon County. The terror of the Molly Maguires was such in the county [Schuylkill] that none felt secure who condemned their proceedings." Thomas J. Foster, through his *Herald* newspaper, led the attack condemning the Molly Maguires.

Convictions and executions of alleged Molly Maguires began in 1877 and continued until October 1879. But in the meantime, the costs associated with hiring extra police protection for the newspaper and its publishers "during the unearthing of the most fiendish conspiracy in the history of the State [Pennsylvania]" forced Foster and Boyer into a Sheriff's Sale of the *Herald*. Afterwards, the publishers were able to reopen, printing the *Weekly Herald* and the *Mining Herald* - until November 12, 1883 when the building and its contents were destroyed by fire.

Somewhere during this trying time, Foster hit on the idea of educating miners and those in charge of mines "in the theory and and scientific principles involved in their various duties, not only for the benefit of the individual, but for the protection and safety of human life, by familiarizing them with the conditions and dangers of the mines." The *Mining Manual* and *Mining Herald* were Foster's first such accomplishments in 1881.

After the fire, Foster began to look elsewhere, and settled on Scranton. By 1890, The Colliery Engineer Company had been formed (with his first cousin Rufus J. Foster), and by 1895 was expanded into the International Correspondence Schools (ICS). The International Textbook Company was organized to provide textbooks to be used in conjunction with the ICS course materials, which were mailed to students located all over the world. Rufus Foster had been an engineer with the Philadelphia & Reading Coal & Iron Company before joining his cousin in Scranton.*Today*, from its headquarters, still in Scranton, ICS continues to educate those around the world who cannot physically access a regular classroom.

Chapter 55

The exception was salt - white gold.

For the men who had accepted the responsibility to carry the legacy forward, the greatest attraction was still for natural resources, now expanded to include a wider variety. There was a change, however; the new predilection was for those natural resources that were renewable.

That is, with one exception.

The exception was salt - white gold.

Salt is constantly being formed as long as water runs over rocks containing chlorides and compounds of sodium. The mineral halite, however, commonly known as "rock salt", is found underground in beds deposited by the dehydration of ancient bodies of salt water. Halite is mined; salt is precipitated.

From prehistoric times, there was no substitute for salt. Salt was a vital food commodity: it was used for preserving food, and for improving the taste of food. (*Today,* uses have expanded and salt is also the source of chlorine and sodium compounds such as baking soda.) In the past, salt was so valuable that it was almost never bartered, but could be exchanged only for cash, gold or silver, even in the early days of the new United States. Settlers in northeastern Pennsylvania knew that "salt could not be had in exchange for grain or dairy products. That [salt] was a cash article."

Historically, salt had been the catalyst that opened up "Black Africa" countries to Portuguese explorers, thus beginning the Age of Discovery. In Africa, south of the Atlas Mountains of Morocco rich with salt deposits, lies the Sahara Desert. On the southern rim of the Sahara, the Kingdom of Mali was rich with gold - but had no salt. Caravans trekking from the north to Mali with salt were welcomed, and returned to Moroccan trade centers enriched with an inordinate amount of gold. One particular King of Mali, Musa I, became curious about the world the traders came from with their supplies of salt, and, having become a convert to Islam, set out on a pilgrimage to Mecca in the years 1324-1325 A.D. Like a true first class traveler, Musa traveled with lots of money and a retinue, and took the long way around, stopping at "tourist attractions" along the African shore of the Mediterranean Sea. The legend has continued to this day that the amount of gold Musa dispensed in Cairo, Egypt alone caused the gold market there to be depressed for a generation (approximately 20 years). Because of Musa and his pilgrimage, West Africa quickly gained the reputation of being a source of gold. As soon thereafter as Europeans could build ships and determine a simple form of navigation that would assure them a return trip home, the Portuguese, starting in 1415, were the first to be off for the fabled Gold Coast of Africa (*Today*, in addition to Mali, the countries of Guinea, Sierra Leone, Liberia, Ivory Coast, Ghana, Togo, Burkina Faso, Benin, and Nigeria).

As for the American colonists, from the landing at Jamestown, Virginia in 1607 to the end of the American Revolution in 1783, there had been no thought that white gold would be one of the riches of North American lands. But it was. Salt deposits and the salty water of Lake Onondaga, held secret by the heart of the

Iroquois Nation, the Onondagans at today's Syracuse, New York, became the first discovery of commercially viable salt in the United States. Prior to the 1780s, the few white travelers to the Council Fire kept by the Onondagans had always been cautioned against drinking from the water of Lake Onondaga because it was "bad." It had not been until after the American Revolution and settlers had moved into Indian lands that salt was discovered. (Blossburg coal brought out of Tioga County, Pennsylvania, by Corning Company activities and its successors, had been shipped north to Syracuse almost exclusively to heat the huge vats for evaporating the water of Lake Onondaga to obtain the salt residue.)

For geologists, the science of describing nature is physiography. By physiography, planet earth can be described the same way one would give a physical description of a person. Facial features such as a long, thin face with brown eyes, aquiline nose and curly blonde hair become provinces and sections for the geographer. A province is the general appearance of the land, comparable to the general description of a long, thin face mentioned above. Sections would correspond to details within a province just as eyes and nose are details within a face.

There are seven distinct provinces in Pennsylvania. Most of the anthracite coal is found in the Valley and Ridge Province - Appalachian Mountain Section. To the immediate west of Valley and Ridge is the Appalachian Plateaus Province. The Appalachian Plateaus includes almost all of western Pennsylvania and curves across the top of the state to also include much of the land that was in the original Connecticut claim, such as Tioga, Lycoming, Bradford, Sullivan, Susquehanna, Wyoming, Wayne, and Pike Counties, and portions of Luzerne, Lackawanna, Carbon, and Monroe Counties. The Appalachian Plateaus Province is a land characterized by high, flat-topped divides separated by steep-sided valleys in which deeply entrenched streams flow. The majority of Pennsylvania's soft coal bituminous fields, including

Blossburg coal, lie within the Appalachian Plateaus Province. So does a goodly portion of central and western New York State.

The salt of Syracuse, New York, is just over the Plateau boundary line in the Eastern Great Lakes Lowland Province.

By 1886, mining expert George Filer, a partner on some of William Winton's projects, had confirmed the presence of salt deposits, again just off the rim of the Appalachian Plateau, in Livingston County, New York.

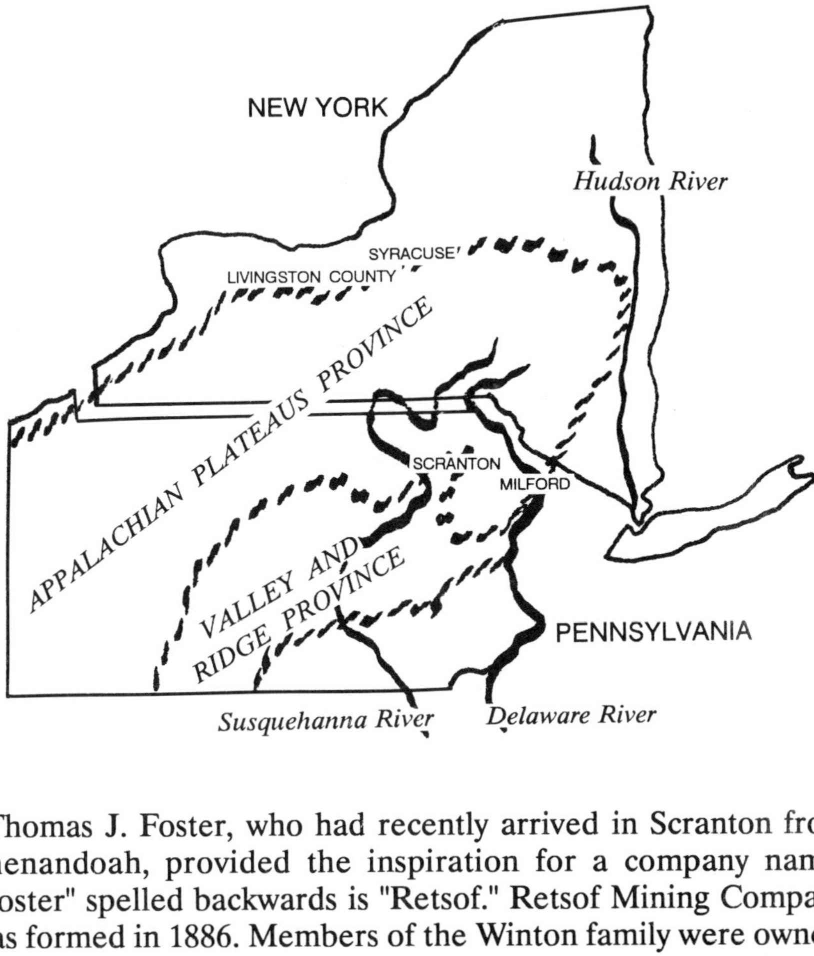

Thomas J. Foster, who had recently arrived in Scranton from Shenandoah, provided the inspiration for a company name; "Foster" spelled backwards is "Retsof." Retsof Mining Company was formed in 1886. Members of the Winton family were owners of a goodly number of its first mortgage bonds and shares of stock.

Byron Winton's brother-in-law Edward Laton Fuller was president. (Fuller was also brother-in-law to Catherine Scranton, the second wife of brother Charles R. Fuller.) Retsof moved into Livingston County, New York, and began salt mining.

In 1890, another salt company was formed, The Greigsville Salt & Mining Company, also to mine in Livingston County, adjoining Retsof. Members of the Winton Family held stock in Greigsville, too. By the beginning of 1895, Quincy W. Wellington, of Corning, New York, (name sound familiar?) was trustee for many of the Greigsville shareholders in addition to the Wintons: Harry C. Heermans and his family which included John H. Fellows, James Everhart, the Law-Bryden Family, the Thomas Waddell Family, Arnot family members, and Jonathan Wilson Peck. Waddell, from Pittson, was mining at Winton under a lease arrangement, and Peck, from the Scranton area, was interested in lumber.

Retsof acquired Greigsville's 400 acres in Livingston County while Archibald F. Law was president of Greigsville in 1900. Fuller remained as head of Retsof, which under his guidance turned into the International Salt Company. *Today*, we know Retsof as Akzo Nobel Salt, Inc. and as Sterling Table Salt, all part of the Akzo Nobel international corporation. In March 1994, after the snowiest winter on record for northeastern Pennsylvania, a portion of Akzo's mine at Retsof, Livingston County, New York began to flood. A March 12, 1994 earthquake at Cuylerville, also in Livingston County, which measured 3.8 on the Richter scale, was blamed for the rock fissure that was releasing water from an aquifer at the rate of 14,000 gallons per minute (gpm). Akzo expects to lose a portion of its deep halite mine as a result of the flooding.

When Fuller died in 1909, his son Mortimer Bartine Fuller succeeded as president of International Salt. (To see what a young Mortimer Fuller was doing on New Year's Day of 1886, see the page from seven-year old Elsbeth Winton's Diary - attending a matinee performance of a play. Mortimer was ten at the time. *Top illustration on page 271.*)

Chapter 56

- the green gold of the New World.

One of the new banks that started up in Scranton in the 1880s was Lackawanna Trust & Safe Deposit Company. Lackawanna Trust had an interesting background, in that while it was new in 1887 when it incorporated, it could at the same time be termed one of the oldest Scranton institutions. George Sanderson & Co. had been established in Scranton in November 1855. Years later, in August 1873, Sanderson opened Lackawanna Valley Bank with his brother-in-law George S. Kingsbury, his son J. Gardner Sanderson, Charles duPont Breck and Dr. R.A. Samies on its board of directors. But in 1881, with the Scranton Trust Co. & Savings Bank and the Winton banks shut down, John Jermyn, Edward B. Sturges and Charles P. Matthews joined the board in what has sometimes been referred to as "a reorganization" of the Lackawanna Valley. The other 1881 change involved Sanderson's son stepping down as vice-president, to be replaced by Robert Thompson Black from the Wilkes-Barre area. Black had often worked with John S. Law, and had participated in establishing the Second National Bank of Wilkes-Barre. With the "reorganization," Lackawanna Valley became a chrysalid for Lackawanna Trust in 1887.

One of Lackawanna Trust's earliest activities after it was incorporated was support for expansion into one of the most renewable of natural resources - lumber and timbering.

When William Jessup and Michael Meylert were working on Lackawanna RailRoad and its coal mine at Jessup in the mid-1850s, adjoining land owner George M. Hollenback supplied much of the timber that was needed. According to ledger sheets kept by Meylert and Christopher Ward at the time, some of the wood was processed at Peck's mill.

The Lackawanna River follows a course like a reverse 'S' along one edge of today's Jessup Borough. On the protrusion created by the outward curve of the 'S,' Samuel Peck, from Massachusetts settled about 1831. (Peck is not a direct descendant of The Reverend George Peck.) At the time, the area from Providence Township to Carbondale, designated Blakely Township, was primarily a wilderness. Peck's main support for his family was in timbering or logging, and from the saw mill he built. To this day, the area where he settled is known as Peckville.

Two of Samuel Peck's children, John Dwight Peck and Jonathan Wilson Peck, established their own lumber business in 1861. Peck Brothers turned into Peck Lumber Manufacturing Company. After Jonathan Wilson Peck's sons Fenwick L. Peck and Edson S. Peck came of age, the lumber business became J.W. Peck & Son, retaining this name until 1886. From 1875 to 1881, Fenwick Peck was in charge of timbering the Dunning tract (A.B. Dunning) at Dunning. But in 1881, Mutual Life Insurance Co. of New York filed suit to collect $9,000 it claimed was owed by Dunning, and by William W. Winton as guarantor of Dunning's note. (A legal battle followed when the insurance company seized Dunning's property in Dunning and continued to press for the full amount from Winton, even chasing after Winton's West Virginia properties.) Leaving the Dunning tract, the Pecks moved closer to Scranton, into Springbrook, and timbered there until 1886.

When Fenwick Peck located over 3,000 acres of prime hemlock near Mina in Potter County, Pennsylvania that could be had for $60,000, a decision was made in September 1887 to incorporate Lackawanna Lumber Company. (Mina is about five miles west of Coudersport, the Potter county seat, on Route U.S. 6.)

William E. Dodge, by himself and with his New York City business partner, Anson G. Phelps (also Dodge's father-in-law) trading as Phelps, Dodge & Co., acquired extensive timber lands in Pennsylvania at the end of the 1840s and the beginning of the 1850s. In Tioga County alone, Phelps & Dodge operated on 25,000 acres. But the lumber firm of Fox, Weston & Bronson, sited at the confluence of the Tioga, Conhocton and Canisteo Rivers in Steuben County, New York (near Corning) had a voracious appetite, and most of Tioga's trees traveled north. By 1871, Phelps & Dodge had decided to headquarter at Williamsport, in Lycoming County, directly south of Tioga; their saw mills, etc. "were dismantled and removed to Williamsport, where still better advantages were secured for the manufacture of lumber."

Much of Pennsylvania's forest north and west of Lycoming County was open to commercial enterprises and exploitation. Often individual investors from outside Pennsylvania held title to some of the timber tracts. One such investor was Robert Looney, a feed and grain merchant and saw mill operator at today's Wende, a small community about 15 miles due east of the Buffalo city limits in New York State. Looney owned almost 2,000 acres near Liberty in McKean County, just over the border with Potter County.

From Groton, Tompkins County, about 15 miles west of Cortland, New York, two brothers, Charles W. Goodyear and Frank H. Goodyear, left for Buffalo, New York (Erie County) in the late 1860s. (This Charles Goodyear is not to be confused with the Charles Goodyear who was born in 1800, died in 1860, and revolutionized the rubber industry by discovering "vulcanization," the process combining rubber and sulphur to produce the kind of rubber manufactured today.) Charles Goodyear of Groton was the older of the two brothers and to study law departed first for Buffalo, where he later joined Grover Cleveland's law firm. (Cleveland served as U.S. president twice, bracketing Benjamin Harrison's term of office.)

Frank Goodyear, following Charles to Buffalo, got only as far Wende, then called Looneyville. There Frank worked for Looney, married his daughter, and became inspired with Looney's investment in "Penn's Woods." Frank and his father Bradley Goodyear, with two others, Daniel Bailey and George Briggs, were able to buy the land at Liberty, Pennsylvania from his father-in-law's estate in 1873. By the end of 1873, Charles Goodyear had joined the lumber venture, and an additional 1,772 acres had been acquired at Liberty. In 1884, having operated successfully since 1877, Frank Goodyear added almost 14,000 more acres extending eastward into Potter County.The Goodyears' method of operation was to own the timberland outright, in fee, and to sell the sawed, dressed lumber. The actual work at the saw mill and the mill itself, however, were operated by others in a contractual relationship with the Goodyears, at a fixed rate.

Throughout the 1800s, in ever larger increments, the demand for lumber had been increasing. Two kinds of lumber were particularly sought after: white pine and hemlock. Pennsylvania had both, and in abundant quantities - the green gold of the New World.

The pine was used in construction, including shipbuilding, and the hemlock was sought more for its bark than for its wood. The bark was necessary to tan leather. Along both sides of the eastern Continental Divide (the top of the Appalachian Mountain chain extending from Georgia to Maine), pine and hemlock were considered to be of the best quality and were cut or stripped first. A portion of the Divide traverses McKean County. The best lumber was usually found in out of the way places and at relatively high elevations.

So, an integral part of large-scale timbering had to be railroading. As did coal operators, whose railroad tracks laid almost to mine entrances, with coal cars, yard engines, etc. standing by on sidings, lumbermen had to do the same thing. The big difference was that railroad companies undertook the responsibility of reaching mining sites on their own, whereas lumbermen had to be their own railroaders. Freight profits could be high when carrying coal, and there could be an exclusive right of access to coal coming from a specific mine opening into a particular vein of coal; both were sufficient reasons for a railroad to serve the coal operator. These motives were lacking when dealing with loggers.

The Goodyears soon found themselves in the railroad business, so that they could bring out logs in any great quantity from the two faces, east and west, of the eastern Continental Divide. Logging railroads, by reason of their function, were usually ephemeral; tracks were laid down and later taken up, as loggers moved from one portion of an area they were cutting to another. But regardless of how transient railways were in the woods, metal rails, switches, flat-bed rail cars, and one or more engines represented a hefty investment. The Goodyears faced a $90,000 such investment after adding the approximately 14,000 acres to their holdings. In January 1886, Frank Goodyear formed The United Lumber Company to control timber prices, especially that of hemlock. He apparently was worried that his company could not withstand a depressed lumber economy combined with all the additional investments it had had to make in railroad equipment. Independent lumbermen, as members of United Lumber, would hold to prices if need be. Frank Goodyear was president; Judson J.Newman, a dealer in lumber and lumber products at Buffalo, New York served as secretary. About the same time Newman took on his job with United Lumber, he came to Potter County, Pennsylvania, and established a saw mill operation at Sweden Valley, about four miles east of the county seat at Coudersport.

Lackawanna Lumber Company moved in the following year, setting up lumber operations a few miles west of Coudersport. J.W. Peck and his two sons Fenwick and Edson, along with William T. Smith (who was one of the incorporators of Lackawanna Trust, and president in 1887), James L. Connell, his father William Connell, his business partner Isaac F. Megargel, Joseph H. Steell, and Henry Belin, Jr. were some of the incorporators. James Connell was the treasurer of Lackawanna Lumber. Lackawanna Lumber's initial 3,000 acres soon grew to 15,000 acres, plus ownership of just the timber on another 10,000. Within ten years (by 1897) Lackawanna Lumber and F.H. & C.W. Goodyear were producing almost 50% of all lumber leaving Potter County.

Meanwhile, back in Scranton, one success was sparking another and another. Over 16,000 acres in Beaufort and Hyde Counties, North Carolina were purchased in 1889 by the Scranton & North Carolina Lumber Company, also incorporated in 1889. Arthur D. Dean, William Winton's nephew and son of the Isaac Dean who had been active with Winton in setting up banks in the 1860s, was an active member of Scranton & North Carolina's board of directors; J.M. Rhodes was president and Pat Mulherrin was secretary. Initial plans to set up a mill and begin lumber operations had started in the early part of 1888. The acreage was located on North Carolina's mainland, protected by Cape Hatteras.

Another company, the Allegheny Mining and Improvement Company, was actually incorporated first (March 1887), even before Lackawanna Lumber, However, Allegheny did not reveal too much about its activities until two years later, when it changed its name to The Allegheny Company, and had Isaac F. Megargel for president. Megargel would be succeeded by Joseph H. Steell.

And what was Allegheny up to?

Well, between 1892 and 1899, it was acquiring over 200,000 acres in Beaufort and Hyde Counties, North Carolina.

Allegheny's initial land purchases had included a 100,000 acre tract. Later purchases, around 1898, had been to acquire land from two other lumber companies operating in the same area. Scranton Lumber Company and Bell Haven Lumber Company began their corporate careers in 1899 with Joseph Steell as president and James S. McAnulty as treasurer (note that McAnulty, in addition to being William Connell's son-in-law, managed Horatio Pierce's business interest until Pierce's grandson came of age).

In the western part of North Carolina, in Caldwell County, yet another Scranton firm, The Wilson Lumber Company, was operating on 12,000 acres. In 1895, it changed its name to The Wilson Lumber & Milling Company. Charles S. Weston, Horace E. Hand, and William L. Connell were among the men involved. To accommodate the Wilson Company, Lackawanna Trust and Safe Deposit Company opened a branch office in Lenoir, the Caldwell County seat. This was not Lackawanna Trust's only involvement with the various lumber companies tied to Scranton. It provided funds, loans, and mortgage money. Setting up lumber companies was an expensive business. Besides the saw mill(s), every lumber venture required a logging railroad of some length; and as with lumber shipments from Beaufort and Hyde Counties, it needed a boat line with barges going north to the Baltimore, Maryland and/or Washington, D.C. markets. In the beginning, money would also be needed to set up a post office, a company store, possibly a school house, and for generating electricity. *Today,* in Beaufort and Hyde, the Weyerhaeuser Company conducts scientifically managed tree harvesting and reforestation programs.

So much for North Carolina timber. Are there trees anywhere else?

Well, yes - try Mississippi.

The opportunity for the boys from Scranton to extend deeper into the south came after Judson Newman was able to establish a small lumber operation at Hattiesburg, Mississippi by 1895. From Hattiesburg west through Sumrall, crossing the Pearl River to Bude, Mississippi, the J.J. Newman Lumber Company would eventually own over 300,000 acres of southern yellow pine. (Newman himself, when timbering at Sweden Valley in Potter County, Pennsylvania, had owned a modest 3,000 acres.) The vast supply of yellow pine in Mississippi was a welcome addition to the mix of white pine, walnut, oak, hickory, birch, ash and chestnut trees found in Caldwell County and the additional varieties of cypress, gum, maple, holly, and cedar trees found in Beaufort and Hyde Counties, all in North Carolina. A railroad across Mississippi was begun initially by laying about 150 miles of direct track to connect the towns where the saw mills were located. The name of the railroad was Mississippi Central Railroad Company. Mississippi Central first appeared as a reorganization of Pearl & Leaf River R.R. yet Pearl & Leaf was only one month older than Mississippi - it existed from November 20, 1897 to December 21, 1897. More tracks to Natchez on the Mississippi River, and to the Gulf of Mexico seaport at Mobile, Alabama, would bring the Mississippi Central up to 250 miles of main track and almost 30 miles of yard tracks and sidings.

J.J. Newman became the largest timber operation within the interior of the State of Mississippi. After almost a quarter century of tree harvesting, it was estimated that there were still eight billion feet of standing timber in 1923.

With all the good lumber available in Mississippi and North Carolina, it wasn't long before Charles Hebard was enticed from his northern timberlands in Michigan's Upper Peninsula to expand his operations elsewhere. Soon, the Hebard Cypress Company was busy timbering, too. (Hebard had gone out to Ontonagon County, Michigan, with Edward Dolph, Elisha Phinney, and William Winton in the 1870s.)

The Goodyear brothers, left behind in Pennsylvania's Potter and McKean Counties, had expanded somewhat (coal and iron between Pittsburgh and Buffalo, New York) but never to the extent of the Peck/Newman business in Mississippi. Then in 1902, Frank Goodyear as president and Charles Goodyear as vice-president, with others, started up the Great Southern Lumber Company. It outdid J.J. Newman; it had 350,000 acres mostly in Louisiana north from Lake Pontchartrain, with the rest of its acreage lying in Mississippi. (New Orleans touches the southern shore of the lake.) At a point further south than Newman lands on the Pearl River, where the river forms a portion of the border line between

Louisiana and Mississippi, the Goodyears designated an area suitable for building a town of 10,000 as "Bogalusa."

The Goodyears worked out a plan to make Bogalusa a permanent community, and succeeded. Today, the population is stable, at around 16,000. To turn a transient lumbertown into a lasting enterprise, the Goodyears convinced Louisiana legislators that reforestation with accompanying delayed taxation was the best possible alternative for cutover land. In the interim, until new growth could be harvested (from ten to a 100 or more years), the Bogalusa Paper Company was started up. As time went on, Bogalusa Paper was merged with Gaylord Paper Company, of St. Louis, Missouri. After World War II, in the 1950s, the merged company became a part of the Crown Zellerbach Corporation.

Fenwick Peck served on the board of directors of Great Southern Lumber Company.

Was there lumber anywhere else?

What about West Virginia?

The "boys from Scranton" had been in West Virginia since the 1850s when it was the western part of Virginia. In the last few years of the 19th century, they added major timbering activites to their other interests. Fenwick Peck provided the expertise. At the junction of the north and south forks of the Cherry River, about 75 miles east of Charleston, the Cherry River Boom and Lumber Company [CRB&L] located a saw mill at Richmond in Nicholas County. Eventually, CRB&L would own 200,000 acres spread over Greenbrier, Webster and Pocahontas Counties in addition to Nicholas. Its railroad, built to bring out timber from the West Virginia hills, would total over 75 miles, and 16 engines would be needed. CRB&L would also establish saw mills at Camden-On-Gauley, Webster County, and a second mill in Nicholas County at Holcomb. Starting in 1910, Nicholas County would be recognized as the center of the hardwood industry in West Virginia, thanks in part to CRB&L. (In the period from 1879 to 1912, lumbering companies operating in West Virginia managed to timber over 85% of the state's woodlands.)

Peck left CRB&L soon after getting it started. (CRB&L was incorporated in 1900, concurrently with the plans and work in New Mexico and Mississippi.) He had other demands on his time and money. Scranton ties to CRB&L remained, however, and the company managed to survive the "great depression" of the 1930s, partly by concentrating on its paper mill and partly by selling 153,000 acres to the federal government for $20 million under President Franklin D. Roosevelt's emergency discretionary fund. The land that was sold was primarily in Greenbrier County, and is now a part of Monongahela National Forest.

Would there be lumber anywhere else?

Try southern New Mexico, just to the east of Alamogordo.

Alamogordo Lumber Company began its corporate existence in May 1898. Fenwick Peck was president. Like the other logging companies, this one was tied to a railroad - the El Paso & Northeastern R.R. The main reason for the existence of Alamogordo Lumber was the production of railroad ties needed for New Mexico's railroad trackage.

Interest in New Mexico went as far back as the early 1870s, when Robert M. Kidder acquired land at Silver City for C.P. Kidder, Sheldon Reynolds, A.J. Davis, John Wood and J.D. Hoyt, all of Wilkes-Barre. (Another member of the Kidder Family, Luther Kidder, had acted as trustee for Henry Heerman's bequest to his second wife Sarah Ann, when she was remarried in 1845 to Joseph Steele, Jr. William Winton was executor of Heerman's estate at the time.) In 1880 Edward Dolph's son, Edward S. Dolph, worked for a year in the office of the surveyor-general of New Mexico at Santa Fé.

The interest shown in New Mexico was for a very specific reason. Scientific testing in 1872 had determined the presence there of anthracite coal, the only anthracite west of the Mississippi River. Anthracite was found south of Santa Fé, and semi-anthracite at Silver City. Elsewhere throughout New Mexico, deposits of bituminous (soft) coal had also been located, starting at the Colorado border near Raton. (This was in addition to gold, silver, etc.) So the interest in New Mexico for the boys from Scranton was coal, just as it was in West Virginia. And the problem was the same that had originally confronted coal owners in northeastern Pennsylvania - how to get it to market? The problem was solved in the same way it had been solved by the Scranton brothers in the early 1850s - by large financial commitments and railroad transportation. By the 1890s, however, the boys from Scranton would not be bringing in too many outside investors. They were pretty adept at building railroads themselves. At first, railroad ties would have to be brought from east Texas. But as soon as the Alamogordo Lumber Co. could be formed, with Peck in charge, ties could be manufactured in New Mexico in sufficient quantities for railroad tracks to reach all the important mining locations throughout the state.

By today, a legend has grown up around the beginnings of the railroad to Alamogordo and northward. In the 1890s, a developer operating in the Pecos Valley in southeastern New Mexico conceived the idea of a railroad from El Paso, Texas, north across New Mexico to connect with the Chicago, Rock Island & Pacific R.R. at Liberal, Kansas. His name was Charles Eddy; and the only backers he could find for his dream were the boys from Scranton. Word of their plans leaked out, and a group of men, as incorporators of a White Oaks Railway, appeared before El Paso's governing body requesting a franchise for their railroad that would use the same course Eddy wanted.

But word can leak two ways, and by a reverse leak, Clarence Simpson and the other boys from Scranton had their attorney, "the famed William Ashton Hawkins," before El Paso officials demanding that a $10,000 performance bond be posted by the White Oaks group. No such performance bond was forthcoming, and in the spring of 1897, El Paso and Northeastern Railroad/Railway Company was incorporated in Texas and in New Mexico.

By June 1898, the El Paso & Northeastern had reached a point approximately 85 miles north of El Paso. There, on what had once been the Oliver Lee ranch, town lots were marked off for a new community, called Alamogordo. Alamogordo served as headquarters for the El Paso & Northeastern. Immediately east of the new town lay the forests of the Sacramento Mountains, where Peck and his loggers would bring out almost all the timber necessary for building railroad trackage in New Mexico.

Today, Sacramento Mountains are a part of Lincoln National Forest. Immediately northwest of Alamogordo, the world's first atomic bomb was exploded on July 16, 1945, at the White Sands Missile Range's Trinity Site.

Chapter 57

For all their enthusiasm for renewable resources, mining was what the boys from Scranton knew best.

While the Lackawanna Trust & Safe Deposit Company was providing backup for many of the lumber activities, The Traders National Bank of Scranton was handling more of the coal mining business.

For all their enthusiasm for renewable resources, mining was what the boys from Scranton knew best.

Across the United States, it was generally conceded among mining men that "certain Scranton, Pennsylvania investors" were men who had an inordinate amount of knowledge and expertise for developing coal properties.

The railroad that was built north from El Paso, Texas into New Mexico had been designated a "lumber road," but logging was not the only interest of the boys from Scranton. So, the El Paso and Northeastern Railway hastened north from Alamogordo - under a variety of names. For part of the distance, it was still the El Paso

& Northeastern, but every turn and major shift in direction seemed to be sufficient cause to create a new corporate name. After reaching the small coal field around Carrizozo and Capitan and before continuing further north to Torrence, it took a new name to branch eastward; the El Paso & Rock Island Company was formed to extend rail trackage 128 miles northeast to Santa Rosa - and a junction with the Rock Island R.R. Once this was accomplished, travel from Chicago, Illinois to El Paso, Texas would be possible. Meanwhile, the El Paso & Northeastern continued on to Torrence. There at Torrence, another railway, the Santa Fé Central, began its track north through the Estancia Valley to Santa Fé, passing by the anthracite coal field. As a byproduct, produce, livestock, and salt, as freight from along the length of Estancia Valley, became big-profit items. The Santa Fé track was opened in 1903.

Meanwhile, El Paso & Northeastern headed for the Colorado border. Attorney William Ashton Hawkins once again showed his legal ability, as he had in El Paso before El Paso City officials; he examined John Dawson's title to the property that is known as the Maxwell Land Grant (a Spanish grant totaling approximately 1.7 million acres originally) and rendered the opinion to the boys from Scranton that it would hold up in court. The title did hold up in court, and the El Paso & Northeastern was on its way to Dawson, about 35 miles southwest of Raton and the Raton Pass into Colorado. For the last 132 miles of El Paso & Northeastern's track, it was known as Dawson Railway. Leaving Santa Rosa on borrowed rails (from Rock Island) for a 60 mile trip east, to Tucumcari, Dawson Railway began at Tucumcari and terminated at Dawson. This last link was completed by 1903.

Of all the various railroad companies connected to El Paso & Northeastern, the first "spur", also completed by 1903, is probably the most romantic and dramatic. At Alamogordo, the Alamogordo & Sacramento Mountain Railway (A&SM) rose almost 5,000 feet in elevation over just 32 miles, earning for the railway the nickname of the "Cloud-Climbing Route." Six miles from the end of the route, at Cloudcroft (elevation 8,627 feet), a resort hotel and community was developed by the railroad. Part of the ride to and from Cloudcroft involved a switchback. The A&SM is considered one of the legendary railroads of the west.

The El Paso & Northeastern (under all its names), which crisscrossed New Mexico from its northern to southern borders, was an important transportation link within New Mexico. It carried people, food products and cattle in addition to lumber. But more importantly, it fulfilled the purpose of its creators - it provided transportation for almost all of the major coal mining areas as well as connecting them together. Under "Scranton boy" Clarence Simpson, the New Mexico Railroad & Coal Co. met the needs of the 25,000 acres of coal land the boys from Scranton owned. And coal was available for railroads and their steam engines, for domestic uses, and for mining operations, where coal or coke was needed for smelting copper and other ores. For smelting, coal was shipped west to Douglas and Morenci, Arizona and south to El Paso, Texas.

There was still one more area where commercial timber stands were plentiful in the southwest. In eastern Texas, in what is known today as the Piney Woods area, members of the Kirby Family of Wilkes-Barre, Frederick M. Kirby and J. H. Kirby, acquired 800,000 acres of yellow pine trees from Houston Oil Company. The Kirby Lumber Co. and an accompanying railroad were set up in 1901. After 1920, Kirby Lumber would begin a merger with Southwestern Lumber Co., the name of the firm that had come to stand in the place of Alamogordo Lumber Co. *Today,* almost all of the Kirby Lumber trees in Texas' Piney Woods are under federal government control as the Big Thicket National Preserve, and the Angeline, Davy Crockett, Sabine, and Sam Houston National Forests.

While Simpson was administering Scranton affairs in New Mexico, Edward S. Jones had gone off to West Virginia. (Jones' father Edward Jones and Clarence Simpson's father William

Simpson had mined together in the Archbald and Winton areas of northeastern Pennsylvania as Jones, Simpson & Co.)

Almost all of West Virginia had coal. It was bituminous, soft coal, and by the 1890s it was supplanting anthracite in the iron and steel mills. Almost everyone in North America prior to the American Revolution in 1776 had known about the coal in western Virginia. Transporting the mined coal from out of western Virginia's hills, however, had seemed an almost insurmountable problem.

George Washington had worked on the problem of transportation for western Virginia's coal in 1785 as president of The Potomac Company, which was to build a canal to go as far as Harper's Ferry on the Potomac River. To reach the hills of western Virginia, Washington assured his company of the State of Maryland's cooperation by the Mount Vernon Compact. But the project was too big a task for that time, and The Potomac Co. lapsed into silence until later, in 1828, when it provided rights-of-way for the Chesapeake & Ohio Canal Company. Minimum coal production followed, and then came the war years of 1812-1815. It had been western Virginia's mines that had supplied the United States with coal at the beginning of the British embargo in the War of 1812 (between the U.S. and England). But again, the difficulty in bringing the coal to market in commercial quantities from the mountainous area where it was found prevented bituminous coal from taking over the coal market in Britain's absence. The high cost of western Virginia coal opened the way for Pennsylvania anthracite.

Little had changed by 1861, when The War Between the States started. On June 20, 1863, western Virginia became the state of West Virginia. After the war West Virginia's coal fields and timber lands gradually opened up as transportation companies successfully penetrated the mountains.

It was David Gowen and the Philadelphia & Reading R.R. (Reading R.R.), by organizing "the combination" in 1872 to regulate mining and export quantities in the anthracite fields, that made the biggest impact. With the start of "the combination,"

Reading, Pennsylvania R.R., and Lehigh Valley R.R. men moved into West Virginia and acquired what coal land they could. The same was true for industry leaders in the Pittsburgh area. The key point was not that these men were the first or only ones to improve transportation into the mountains of West Virginia, but rather that they shifted the demand for coal to operate iron and steel mills from anthracite to bituminous coal; bituminous could easily be "coked."

With Andrew Carnegie at Pittsburgh in the forefront, the iron and steel industry rapidly changed between the years 1856 and 1885. In 1856, Charles Bessemer of England discovered the effects of a blast of cold air on a large "ladle" of 15 tons of hot, molten iron ore; in 1861, Charles Siemens from Germany discovered the open hearth process, whereby 15 tons was increased to 50 tons; in 1872 Carnegie began building blast furnaces in Pittsburgh; and in 1885, Carnegie learned how to make Englishman Sidney Gilchrist-Thomas' method for using low grade iron ores in open hearth furnaces work. Soft coal, coked, would make the process work the best. (For coking, coal is set on fire and allowed to burn just to the point where almost all volatile gases have escaped. The fire is then extinguished. The remaining coal is called "coke.") With the Pittsburgh area as the center, bituminous coal and coke were brought north mainly from West Virginia, and much of the iron ore from northern Minnesota, Wisconsin and Michigan south, to make western Pennsylvania the largest steel producing area in the world. (This was in addition to Pennsylvania's own iron and coal reserves.)

Fortunately for the boys from Scranton, they became an integral part of the demand for soft coal and coke. "The combination" that had been formed in 1872 was not going to deal them out after all. Actually, they had had a presence in West Virginia from the days when it was western Virginia - when Joseph Fellows had sent his nephew John Heermans there in 1849. Scranton's presence was certainly stronger when William Winton and Edward Dolph acquired 21,000 acres in 1873. But it wasn't until Fenwick Peck headed up the Cherry River Boom & Lumber Company at the beginning of the 20th century, and Edward S. Jones at the same time got the Blue Creek Coal & Land Co. and Kanawha & West Virginia Railroad Co. going, that the mining community realized just how involved Scranton men were in West Virginia.

Chapter 58

In the new era that was beginning, it would be the impersonal institution that owned the vast tracts of land.

Meanwhile, back at Scranton, there remained the unfinished business of the New York, Susquehanna & Western R.R. (NYS&W). It had yet to come into Scranton on its own tracks.

The agreement between the NYS&W and the Delaware, Lackawanna & Western R.R. (DL&W), concluded in December of 1881, was to last for 99 years. Supposedly, there would never be a reason for NYS&W to extend its own trackage beyond Stroudsburg. But multiple grievances, centered around freight rates, spurred the men behind NYS&W to resume their dormant desire. DL&W rates were allowing 50% of gross receipts to coal operators, at a time when the standard rate was 60%. And you know who the mine operators and owners were: John Jermyn, Horatio S. Pierce, William W. Winton, Edward Dolph, John S. Law, Edward B. Sturges, Orlando S. Johnson, Edward N. Willard, Louis A. Watres, etc. Need anything more be said?

But there was the agreement with the DL&W, which guaranteed the men access to tidewater for their coal. Without it, they could be left with tons of a very desirable product yet nowhere to go with it. It was at this time, from the early 1880s to 1890, that the men devised a plan to surmount their vexations and frustrations. The plan worked so well that it was used again and again, from North Carolina to New Mexico and wherever else they were investing and building.

Whatever the project, it would be done in small units. Each unit would start off autonomously and with a "third party" at the head; the third party was either a friend or local-to-the-project's-area promoter who was given a financial interest in the venture.

Could this be a trust? You build something new by first constructing the parts and then assembling those parts to create a whole, finished product, much like prefab units are assembled to build a house or office building. Would this be the kind of trust that the Sherman Antitrust Act of 1890 and Clayton Act of 1914 were enacted to stop?

On March 8, 1892, the Wilkes-Barre & Eastern Railroad Company was formed. In its general description, the company's railroad would operate between Wilkes-Barre and Stroudsburg. It was a new route, beginning at Kingston and climbing into the Pocono Mountains by crossing over the Panther Creek near Spring Brook. The 187' high trestle, extending 1,730' over the creek, was responsible for the railroad's nickname: The Panther Creek Line. For a while, this trestle was the second highest in Pennsylvania.

It was not until September 1893, one month before the railroad was open for business, that the news was released: "On its completion, the Wilkes-Barre and Eastern shall be operated in connection with the New York, Susquehanna and Western as a continuous line." There was a 99-year agreement in effect. The cost of building the road was set at $7 million.

In the same month that Wilkes-Barre & Eastern incorporated, approximately 100 miles east of Wilkes-Barre the Hudson River

Railroad and Terminal Company started up (March 31, 1892). Hudson Terminal was going to build a railroad terminal on approximately 75 acres, including 1,800 feet of river frontage that it owned at Edgewater, Bergen County, New Jersey. Who, among the coal-carrying railroads, would have dreamed that this new company, located more than ten miles north of their terminals around Hoboken, New Jersey, would be of interest to them? Guess again! (There was a clue, however: Winton's son-in-law, Thomas Livey, had been acquiring property for development in Ridgefield Park, immediately west of the Palisades cliffs, also in Bergen County.)

A year later (April 25, 1893) the NYS&W and Hudson Terminal consolidated. The boys from Scranton could now reach tidewater on their own, free of any "combination." (The NYS&W and the DL&W entered into a legal fight over setting aside their agreement. The court decided in favor of NYS&W.)

After a lapse of a little over three years, the Susquehanna Connecting Railroad was started in December 1896. Susquehanna Connecting had the right to lay tracks from a connection with Wilkes-Barre & Eastern to Winton, in Lackawanna County, northeast of Scranton City. Susquehanna Connecting was built, but only as far as Jermyn's collieries in Old Forge, south of Scranton.

So, in a way, Pierce, Law and Winton had succeeded in bringing in rail transportation unhindered by McGowen's Reading R.R. and "the combination," although not as completely as they had originally planned. It was still necessary to use other railroads' tracks for the coal coming from their mines above Old Forge, before hooking onto Susquehanna Connecting, Wilkes-Barre & Eastern and NYS&W engines.

The execution of the final plans for the NYS&W was carried out by the men who were the inheritors of the legacy. It was at a time of sadness. Horatio S. Pierce had died in 1889. Then, Edward Dolph passed away in April 1890, John S. Law in 1893, William Winton in December 1894, Winton's wife Catherine Heermans

Winton a half-year later in June 1895, and their son A.H. Winton eight months later in February 1896. And George Filer, the mining expert who had traveled so many times to so many places, died in April 1898.

The end of an era had come.

It was the end of another kind of era also, in the sense that it was the end of an age that had accepted the right of the individual to own vast tracts of land in order to put it to use; where the identity of individuals who owned large amounts of real estate could be hidden under a corporation that was to all intents and purposes developing the land for its own money-making endeavors.

In the new era that was beginning, it would be the impersonal institution that owned the vast tracts of land.

The first such institution would be the corporation itself; the concept of the "hidden owner" all but disappeared - WYSIWYG (what you see is what you get) - corporate officers acting in unison. After that, it would be the impersonal banking companies who would be the owners of vast tracts of land.

At the end of the era, after the 1900 census was tabulated, Scranton emerged as the 40th largest city in the United States. What a remarkable growth from 50 years earlier, when a wooded section along the Lackawanna River housed approximately 300 residents.

Scranton was living on its own by 1901, its very reason for existence, the Lackawanna Iron & Coal Co. (LI&C), having departed. In a corporate split in 1881, William W. Scranton, with William Connell, George L. Dickson, Edward P. Kingsbury, Charles E. Judson, Walter Scranton, and George B. Smith, formed a new company, naming it the Lackawanna Steel Co. Scranton's company operated for about seven years, afterwhich it was acquired by LI&C. The merger led to a new name for LI&C, Lackawanna Iron & Steel Co., in 1891. By 1901, company officials had moved the entire operation south of Buffalo, New York, to a place they named Lackawanna, located on Lake Erie. (Lackawanna Steel was the fourth largest steel producer in the United States in 1921, but was operating at a loss. Bethlehem Steel Co., under Charles Schwab, bought Lackawanna and its iron ore deposits.)

The beginning of the 20th Century was a new era in more ways than one.

The passion for "bigger" and the belief that "bigger is better" were philosophies that captivated the business world at the start of the 1900s. The basic popularity for the new economics was that "bigger" was really a way of cutting costs, by approaching the state of ultimate mass production. But when Carnegie Company became U.S. Steel, with 45% of the selling market and involved from mining ore to the finished product, other questions were eventually raised: was U.S. Steel a trust that controlled the market in steel?

The rush for "bigger" gained acceptance with the boys from Scranton. It wasn't long before they, too, had consolidated companies. But for the Scranton men, it wasn't always the issue of buying other companies. In most instances, it was as simple as putting Humpty-Dumpty back together again. All the small companies under which they had been doing business were now brought together under the umbrellas of just a few parent companies. In New Mexico, under Clarence Simpson's direction, New Mexico Central R.R. became the vehicle to bring all of the Scranton-sponsored rail lines in New Mexico together as one in 1908, accounting for 15% of all rail trackage in New Mexico.

In West Virginia, Edward S. Jones worked for consolidation of the coal companies in the New River area, southeast of Charleston. The New River Co. was chartered in 1906, uniting 22 once individual separate mines under one title. It would eventually have a total of 60,000 acres on which to mine for bituminous coal. In a stockholders' suit brought in 1912, the authority of the Scranton interests was challenged. A minority group asked for receivership. Apocryphal tales from that time describe the Pennsylvanians as fleeing back to Pennsylvania on a very long train. But the lower court decision was challenged, and in 1913, nine months after the suit was originally filed, the West Virginia Supreme Court of Appeals ruled in favor of management. So whether the Pennsylvanians had ever left New River territory or not, they were definitely back in town. Down through the years, the New River Co. was recognized as a pioneer in company safety programs, in surface mining, and in land reclamation. Experiments as early as the 1940s determined the most suitable trees and grasses for land restoration. For years, 10,000 tree seedlings were planted annually as part of New River's reforestation program. In recent operations, New River entered into an agreement with a French steel manufacturer to supply bituminous coal.

The New River area had been one of the last areas to develop as a mining community in West Virginia. The reason was not that it wasn't known that coal was there; the problem was lack of transportation. The Chesapeake & Ohio R.R. (C&O) hadn't made it into New River until 1893. The C&O entered West Virginia from the east at White Sulphur Springs. From the northeast, two railroads crossed the Cumberland Gap section of Maryland, the Baltimore & Ohio R.R. (B&O) and Western Maryland RR. It took the Western Maryland until the end of the 1890s to reach Elkins, the county seat of Randolph County, approximately 75 miles into the State of West Virginia. Whatever railroads could get in and out of the many hills of West Virginia stood to make large profits. Western Maryland wanted to be one of those railroads.

In April 1900, at Greigsville, Livingston County, New York, Archibald F. Law, as president of Greigsville Salt & Mining Co., merged Greigsville into Retsof Mining Co. Edward Fuller was still president of Retsof. With almost all of the salt deposit in Livingston County under one name, Fuller moved ahead and transformed Retsof into International Salt Co. in 1901. International included salt production in Kansas and Michigan as well as in New York State. (International Salt would acquire Detroit Rock Salt Co. in 1913.) Fuller was now ready to join his friends in West Virginia.

Western Maryland R.R. had begun its existence in 1853 as the Baltimore, Carroll & Frederick R.R.; a month later on March 21, it became the Western Maryland Rail Road Company. Western Maryland had wanted to go west and it did, meeting the Chesapeake & Ohio Canal at Cumberland 20 years later in 1873. By 1892, Western Maryland was in West Virginia but only to Cherry Run, about 30 miles west of Hagerstown, Maryland. Starting in 1859, the City of Baltimore had invested in almost every mile of track that Western Maryland had laid. At the turn of the century, Baltimore was holding almost $9 million of Western Maryland notes. The City of Baltimore wanted out.

Enter Fuller. The boys from Scranton could really use more railway in West Virginia; if they could also easily reach a market place, it would be even better. George W. Gould, Jay Gould's son, at the same time was working on building a transcontinental rail system from his base, the Missouri Pacific R.R. Everybody would work together. Fuller assembled a syndicate (dubbed the Fuller Syndicate) with himself, Gould's general counsel Winslow S. Pierce, and Myron Herrick, former governor of Ohio and former U.S. Ambassador to France, constituting the driving force. Also in the syndicate were Edwin Gould, president of the St. Louis Southwestern R.R., Howard Gould, Joseph Ramsey, president of the Wabash R.R., A.W. Krech, vice-president of Wheeling & Lake Erie R.R., Louis Fitzgerald, president of Mercantile Trust Co., J.H. Hyde, vice-president of Equitable Life Assurance Co., John T. Terry, John W. Gates, Cornelius Vanderbilt, James Parmelee, A.H. Calef, G.M. Dodge, W.H. Taylor, S.B. Elkins, R.C. Kerens, and Walker Scranton. (William Walker Scranton was half-brother to Joseph Augustine Scranton, both sons of Joseph Hand Scranton.)

In January 1902, the Syndicate acquired the West Virginia Central & Pittsburgh Railroad for $18,082,910.39. A few months later in the summer of 1902, the Fuller Syndicate reached an agreement with the City of Baltimore. For $8,751,370.45, the Syndicate would acquire Baltimore's interest in the Western Maryland R.R., extend the rail line further into West Virginia to connect with the West Virginia Central, and also extend eastward to a tidewater terminal which it would erect in Baltimore harbor. According to tradition, the City of Baltimore was so grateful at getting its $8+ million back that it erected a statue to Western Maryland's retiring president John Mifflin Hood and repaved and rebuilt many city streets.

At Port Covington, on the Middle Branch of the Patapsco River in Baltimore Harbor, the Fuller Syndicate erected an 840' long freight pier. Two sets of railroad tracks ran the length of the pier. It was covered with a steel shed. Another pier, for coal, was built beside it. The coal pier was 729' long. Ocean-going vessels could be handled at either pier; there was a 70' trestle. There was a 600' bulkhead. *Today,* Port Covington backs up to the fabled Inner Harbor at Baltimore. Ft. McHenry, where Francis Scott Key was inspired to write the national anthem *The Star Spangled Banner,* is at the point of land about one and one-quarter miles past Port Covington.

Step by step, the boys from Scranton were moving with the times: West Virginia experienced consolidation; there was consolidation in New Mexico, and in the salt industry. What would happen in the lumber trade?

United States Lumber Company was formed in 1901. It supposedly consisted of just the lumber companies that had been operating in Pennsylvania and Mississippi. But its board of directors included Arthur Dean, Fenwick Peck, Fred Kirby, and newcomers Charles Woolworth, Cyrus Jones, and Edward J. Lynett, thereby showing that everyone was pulling together. As for U.S. Lumber's companion company the Mississippi Central Railroad, it did very little in the way of enlarging. Mississippi Central did absorb Natchez & Eastern Railway in 1909. Mississippi Central continued to operate on its own until 1966, when Illinois Central Railroad Company began the necessary steps to acquire it. Illinois Central and Gulf, Mobile & Ohio merged in 1972 to become Illinois Central Gulf R.R.

(Charles Woolworth opened the fourth 5¢ & 10¢ Woolworth store for his brother Frank Woolworth in Scranton in the 1880s.

He also opened the next three Woolworth stores in Maine and six in New York State. Charles Woolworth made Scranton his home. Cyrus Jones had built the Grand Union Tea Company up to 176 stores nationwide by the early 1900s; and Edward Lynett had become the owner of *The Scranton Times* newspaper in 1895. *Today*, the Lynett Family is still involved with *The Scranton Times.*)

There had been a kind of consolidation of the various mining companies also. In 1899, Archibald Law combined eight of the collieries operating in Lackawanna and Luzerne Counties into Temple Iron Co. Afterwards, Law was in charge of Temple Iron. By the merger, approximately 8,000 people were given employment. (Temple Iron derived its name from an iron mill just north of Reading, Pennsylvania.)

When the consolidations first started to take place, most of the men had been actively associated with William Winton, Horatio S. Pierce and John S. Law. They could not look on the larger business units as anything other than that - just larger versions of how they had always been operating. In many instances, the consolidated company could be considered as public recognition of something the men had always known to be true - that many small companies had always been the parts of a whole.

But two things were occurring after the consolidations. For one, many of the men were themselves now becoming a part of the "older generation" and would soon pass from the scene. Who would step in to continue? Who would carry the legacy - leaders in economic activities - forward?

The concept of the corporate hierarchy had taken hold. It would mean leaders from among the men who had come to work for a company at an early age, and moved up, up "the corporate ladder," that would replace those who had accepted the responsibility to carry the legacy forward. But the replacements were of a different background than the men they were to replace. The next generation to come along would have a fear, one that had not really been known or considered by Winton, Pierce, Sturges, Watres, the Laws, the Connells, Simpson, Peck, Fuller, Kemmerer, Hebard, and all the others who had worked together. Those men had not had the fear that they could lose their jobs; up to the time of the consolidations, they were accustomed to making their jobs happen.

Their replacements, however, had almost always worked in an environment where they could be dismissed or fired at someone else's displeasure. So, the character of the next generation of company officials could not be the same as the men before them.

Then, there was the gradual shift in ownership. Land and personal property such as securities were passing into the hands of banks, by way of their trust departments. A corporate officer's job duties might go to his hand-picked protegé, but his ownership of land, of the company, of controlling shares of railroad stock or whatever would usually pass into an estate, to be held for beneficiaries. Who would know how assets were allocated within a bank's trust department? Various banking regulations were and are put in place to control commingling of funds and other improper banking practices. But the banks were not interested in defrauding beneficiaries. To the contrary, banks were more interested in enlarging their trust departments. For now the bank, that truly impersonal institution, was coming to stand in the place of the large land owner.

The shift to banks as the power base was never perceived by Winton, Pierce, Law, or Fellows. They had been bankers; the men who came after them were bankers, too. These men had not had the fear that corporate employees on the rise experienced. Winton and the others looked on banks as the vehicle that banks had first started out to be, a tool to expedite business matters. The employees who were singled out as protegés could never have that outlook. When it came the protegés' time to fulfill the promise expected by their mentors, the protegés would try. But because of a subservient fear that could never leave them, the protegés would not continue in the style of the men before them. Instead, they would succeed only in attributing to an impersonal institution a human quality - leadership. To bankers, the axiom was now "money is power." Land was no longer a sign of anything. Land to be used was no longer important. The legacy was gone.

By 1940, Fenwick Peck was dead. He had been worth millions of dollars; his companies had always shown a profit. Fenwick Peck died penniless. It was the U.S. Lumber Co. which kept him going near the end, giving him a house and living expenses.

But by 1940, if Peck was not living, who was running U.S. Lumber? If Peck had been depended on U.S. Lumber in the last years of life, who had been in charge of U.S. Lumber to make that decision?

Arthur D. Dean died in 1926. A Winton and Heermans cousin (he was William Winton's nephew), Dean had been a part of U.S. Lumber from the very beginning in 1901. When Dean died, almost two-thirds of the monetary value of his estate came from U.S. Lumber and Mississippi Central Railroad stock. In the first year following his death, his estate took in an additional $19,320 in dividends from just U.S. Lumber. Dean instructed in his will that all of his assets including income/dividends after his death were to go into a trust that was to be administered jointly by Lackawanna Trust Co. and one of his sons. Within two years after his death, Lackawanna Trust and Scranton Trust had merged into the Scranton Lackawanna Trust Company.

Dean had also been carrying a large debt with First National Bank of Scranton, and left instructions that repayment of the loan be balanced against some distribution to his heirs, his children. When the "Great Depression" of 1929 came, Dean's estate did not do too poorly; U.S. Lumber continued to pay sizeable dividends, finally dropping to about $4,000/year in 1936.

It was then that First National collected the remaining balance on the debt. First National claimed all the shares of U.S. Lumber stock in Arthur Dean's estate, stating that depressed prices had made it almost valueless. Who was there to contradict the bank's statement? Certainly not Peck.

But once First National had possession of Dean's U.S. Lumber stock, it announced that it would not sell all of the shares at once, as that would depress the market. Did First National ever sell any shares? World War II was about to start, and the need for lumber would increase dramatically. And when the shares of U.S. Lumber would be sold at varying future intervals, most certainly the shares would be bringing higher prices because of wartime demands. Shouldn't all money received from the sale(s) above the loan amount be returned to the heirs and estate of Arthur D. Dean?

Did anything like that happen? No, of course not. First National continued on its policy of accumulating U.S. Lumber shares where it could. All of the men involved with the lumber company originally had been accustomed to using banks as part of their operating procedures; almost all of the men had helped to start banks. The relationship between Dean, his contemporaries, and the banks they had established, is reflected in part in Dean's will: "I authorize and direct my Executors and Trustees to borrow money from time to time for any purpose in the administration or distribution of my estate, or to continue or renew any loans made by me in my lifetime," and so on. Who, today, would think to leave instructions in their will that banks are to loan money, and expect the instructions to be followed? But Dean did.

When First National called in its loan to Dean, supposedly in order to balance its own books, then sooner or later, First National should be converting the intangible paper asset (stock certificates) into hard currency. But First National didn't do that; instead, it held on to all the U.S. Lumber stock it could get. And there was a time, during World War II war years, when U.S. Lumber, without fanfare or much attention, bounced back and forth between being a Delaware corporation and a Pennsylvania corporation, until 1952 when a 31-year employee of U.S. Lumber, Claire Barrett, and one director, Harry L. Jones, were apparently the only ones living from the 1920s connected with the lumber company. In 1952, the assistant secretary gave testimony that Lackawanna Lumber Co. had been dissolved by court order in October 1945, and that U.S. Lumber, as the sole holder of Lackawanna Lumber's shares of stock, had assumed all assets and liabilities of Lackawanna. As an afterthought, U.S. Lumber's counsel, Hugh J. McMenamin, asked that all references to U.S. Lumber as a Delaware corporation be deleted, as the Pennsylvania and Delaware corporations were one, a Pennsylvania corporation, as of July 7, 1948. McMenamin was a lawyer with the Scranton law firm of Harris, Warren, Hill & Henkleman, "outside" council for First National Bank of Scranton.

For a company whose stock didn't have too much value, that was to have been sold as soon as possible, it was certainly leading an active life. Especially when as a Delaware corporation it had taken title to several tracts of land in Potter County, Pennsylvania from Lackawanna Lumber in 1945.

In 1956, Frank Hemelright was signing as president of U.S. Lumber Co. Hemelright was on his way to becoming president of First National. A very young Frank Hemelright had been an assistant to Archibald F. Law in 1903.

By 1965, another member of First National Bank of Scranton's independent counsel, Edward Warren, was president of U.S. Lumber and ready to do business for the company. (First National had acquired Scranton Lackawanna Trust Co. in 1956, and in 1958 changed its name to Northeastern Bank of Pennsylvania.) Warren collected $500 for U.S. Lumber from another lumber company operating in Potter County, Pennsylvania. By 1965, the law firm was Warren, Hill, Henkleman & McMenamin.

Six years later, in 1972, U.S. Lumber Co. was still going strong. This time it had Joseph C. Kreder as vice-president and Claire Barrett was still secretary. Kreder did not become a partner in Northeastern's independent counsel firm for another two years, after Warren died. Starting in 1974, it was Henkleman, McMenamin, Kreder & O'Connell. By Kreder's signature, U.S. Lumber was merged into Watson Oil Company on March 10, 1972. For handing U.S. Lumber's assets over to Watson Oil, U.S. Lumber stockholders would receive stock in City Investing Company at the rate of approximately one share of City Investing for every two shares of U.S. Lumber.

And just who was City Investing in 1972? Including Watson Oil and its other subsidaries, City Investing listed among its assets: 22,000 acres in Orange County, New York, known as Sterling Forest; 27 acres in Edison Township, Middlesex County, New Jersey; 20% interest in Moorestown Shopping Center, Burlington County, New Jersey; Parke-Bernet Building at 980 Madison Avenue, New York City; 82,000 acre San Cristobal Ranch at Santa Fé, New Mexico; $20 million development Lynwood Estates in Desoto, Texas; $4 million development in Albuquerque, New Mexico; a 750 home development in Colorado Springs; World Color Press, magazine printer; National Bag Co. of Australia; Motel 6, Inc.; Wilson Shipyards and excursion boats; and Southern California Savings & Loan Association. City Investing was also exploring for oil world-wide. City Investing was listed on the New York Stock Exchange with the ticker symbol CNV.

In 1992, CNV was still listed on the New York Stock Exchange but with the company name of Convertible Holdings and described as a dual purpose investment company. CNV had been trading between a low of $6.75 and a high of $7.68 throughout 1992. Based on these prices, and if the number of shares had not changed, whoever had accepted CNV stock for U.S. Lumber Co. shares in 1972 would still have, in 1992, approximately $1.5 million.

The handling of U.S. Lumber Co. stock was not necessarily particular to Arthur Dean and his heirs, or to Peck. Almost all of the heirs of the men who had set out from the center of the United States' first source of commercially viable energy, Scranton, to travel the world serving as economic leaders, had experiences of one type or another along the same line.

In May 1967, Willard Henkleman, another member of the law firm that worked so closely with Northeastern Bank, at that time Warren, Hill, Henkleman & McMenamin, prepared and filed a petition to quiet title a small lot of land originally part of the Pierce, Law and Winton tract in Blakely. In the document he filed with Lackawanna County Common Pleas Court, Henkleman wrote, "The whereabouts of W.H. Winton, Trustee, Defendant in said action, are unknown, and if deceased, the identity and whereabouts of his heirs, successors, devisees and assigns is also unknown to your Petitioners." Yet, only 18 months earlier the same law firm, Warren, Hill, Henkleman & McMenamin, represented two Winton heirs, identifying them as such in a document filed in Lackawanna County. The purpose of the document was to effect a transfer of Winton trust property between the two Winton heirs.

(Very unfortunately, the new age that brought "bigger is better" apparently considered its representatives for quantity and not for quality.)

The French political writer Alexis Charles Henri Maurice Clérel de Tocqueville, known as Alexis de Tocqueville, toured the United States starting in 1831. Afterwards, he wrote an in-depth analysis of the U.S. political system. It was published in France and translated into English between 1835 and 1840. DeTocqueville's *Democracy in America* became a classic and is still referenced in American business circles.

De Tocqueville wrote, "What is most important for democracy is not that great fortunes should not exist, but that great fortunes should not remain in the same hands. In that way they do not form a class."

Did some of the officials of First National Bank/Northeastern believe Joseph Fellows, Horatio S. Pierce, John S. Law, Winton and the others had defied de Tocqueville's observations? Were his observations sufficient justification for an impersonal institution to stand in the place of an individual?

Or would the impersonal institutions - banks - with their money portfolios, loans and determination of credit-worthy profiles by networking with other banks be the trust, the kind of trust the Sherman Antitrust Act of 1890 and Clayton Act of 1914 were enacted to stop?

Chapter 59

After World War II, there was no doubt about it, the United States was truly a world power and a world leader -
The anthracite mines have fallen silent.

One of the men who had worked the closest with the Wintons and the others was Louis Arthur Watres. Besides participating in mining leases, Watres had shown himself adept in finances, and was a popular vote-getter. He was elected a Pennsylvania State Senator in 1882 and 1886 and Lieutenant-Governor in 1890. In 1893, he served as the Pennsylvania commissioner for the Columbian World Exhibition at Chicago, Illinois.

Watres had also developed an interest in hydrology, and had become particularly adept at analyzing water courses and land terrain. He had been an incorporator of numerous potable water supply companies, and Harry Heermans had designed many of the systems.

The boys from Scranton were intrigued with the possibilities of electricity. The times had changed, however, from the days when the Scrantons and William Henry had come to the Lackawanna Valley. Then, what was going on in northeastern Pennsylvania greatly interested New York City investors; that was because of the "black stones that burn." But producing electricity was not limited to just northeastern Pennsylvania. This time around, New York entrepreneurs could succeed by staying in their own backyard, which they did. Men such as Moses Taylor and Samuel Sloan, who had worked so hard for the success of the Lackawanna Iron & Coal Co. and the Delaware, Lackawanna & Western R.R., devoted their energies to turning all of Manhattan Borough, New York, into one electric utility, which they did by 1912 as Consolidated Gas Co. The keen interest of these men in their own home area, instead of in the Scranton area, meant that electricity development in northeastern Pennsylvania would be a "homegrown" product.

For a while, there was great interest in Scranton in converting the mountainous culm piles in the anthracite region into energy for steam heat and/or for generating electricity. (Culm was a waste product in the mining process, usually a mix of some coal and rock matrix. Because it was a waste from mining, the more mining, the bigger the culm piles. No, or very little, vegetation could be induced to grow on the culm, rendering the ever larger "black mountains" perpetually unsightly.) The idea of using water, reverting back to the concept of the waterwheel and mill, to turn turbines for generating electricity, gained popularity, especially as water was the ultimate example of a naturally renewable energy source. So for a while, Watres traveled east when leaving Scranton and wandered throughout the Pocono Mountains.

About 25 miles east of Scranton lay the village of Hawley, at that time a mill town. It was an important stop on the D&H Canal, but in 1898 the D&H closed down its canal permanently. Hawley was also the terminus for the Pennsylvania Coal Co.'s gravity railroad. The community was located at the confluence of the Lackawaxen and Wallenpauack Rivers and Middle Creek. Watres was particularly intrigued with the possibilities he saw in the Wallenpaupack. When he returned to Scranton, plans were formulated. The plans were a long time in taking shape, but then the project that had been planned was monumental.

In 1896, the Hawley Electric Light & Power Co. was formed with William L. Connell as president. (Connell had taken over operation of Green Ridge Coal Co., the Winton mine in Scranton, and was twice mayor of Scranton.) Watres, for the moment, had other activities intervening to claim his attention and time. For one, he joined Edward S. Jones in West Virginia on the New River Co. project. For another, by 1904 he was back in Meylert territory (Sullivan County), helping to start up The Nordmont Chemical Co. for the purpose of manufacturing and selling wood byproducts, and helping the accompanying railroad, Wyoming & Sullivan Railroad Co., to operate from a junction with Williamsport & North Branch R.R. to Ricketts, a station on the Bowman's Creek Branch of Lehigh Valley Railroad, in Wyoming County, a distance of about eight miles. Nordmont had acquired Pentecost Lumber Co.'s interest in 3,500 acres in Davidson Township, and had added another 4,400 acres in Laporte Township, both in Sullivan County.

By 1907, however, Watres was ready to begin work on the new project. As a prelude, he headed up a new trust company, the Scranton Trust Co. (Scranton Trust and Lackawanna Trust merged in 1928 to become Scranton Lackawanna Trust Co.) Orlando S. Johnson, Joseph J. Jermyn, Archibald F. Law, Thomas Watkins, Charles Woolworth, Fred Kirby, Thomas Foster, and Charles Matthews were there with Watres, in addition to several new names.

At the beginning of 1910, plans had been finalized for the new venture. Watres brought his nephew Louis Healy back from East Liverpool, Ohio, just over the borders of both Pennsylvania and West Virginia. Healy had been trained as an electrical engineer, and had become head of the United Power Co. in East Liverpool.

On his return to Scranton, Healy was dispatched to Wayne and Pike Counties. On May 14, 1910, Healy purchased 2,240 acres near the junction of the Wallenpaupack and Lackawaxen Rivers. It was only the beginning.

What had Louis Watres seen on his travels in the Poconos?

Watres saw a long narrow valley, with the "swift and slow" Wallenpaupack River wending its way northeast to meet the Lackawaxen at Hawley. He also saw a drop of almost 400' in elevation as the Wallenpaupack approached Lackawaxen River. Watres saw the potential for a hydro-electric project that could provide electricity to northeastern Pennsylvania. "Mr. Healy's return to Scranton was for the purpose of undertaking the utilization of the Paupack river water power for the generation of electricity to supply the Lackawanna and Wyoming Valleys."

Healy's purchase was only the beginning. Approximately 12,000 acres would be needed to dam the Wallenpaupack back to Ledgedale, a distance of over 13 miles from the dam, which was to be located at what had once been land speculator and Declaration of Independence signer James Wilson's Wilsonville.

Clarence D. Simpson was back in Scranton after his stint getting the New Mexico Railroad & Coal Co. operational. He was available to work with Watres. On July 8, 1910, Watres, with Simpson, Healy, and Lewis and Frederick Stillwell incorporated The Paupack Power Co. Paupack Power had the right to erect reservoirs on the Wallenpaupack and its tributaries between Hawley and Laureldale with a dam and accompanying power plant at Wilsonville. The corporation would issue 4,000 shares of stock for a total value of $400,000. As of July 8, $40,000 had been collected.

A little over nine months later, on April 24, 1911, Wallenpaupack Power Co. was incorporated. It, too, could erect reservoirs on the Wallenpaupack River and its tributaries and build a power plant for generating electricity in Palmyra Township, Pike County. This corporation had a total capital of $5,000. The 50 shares were divided equally between - oh, my! look at the names! - Louis A. Watres, Clarence D. Simpson, Louis Healy, Lewis Stillwell, and

Alvan Markle. $500 had been collected. Three months intervened before yet another corporation was formed. This one, Wayne Development Co., was also capitalized at $5,000. Its purpose was to construct dams/reservoirs, and acquire and clear land necessary to build dams in Pennsylvania. (Yes, you guessed it.) Watres, Simpson, the Stillwell brothers, and Markle were the incorporators. Then there was a lapse of 17 months when not much seemed to be happening. On January 7, 1913, Pennsylvania, New York & New Jersey Power Co. was formed by a merger of the Paupack Power Co. and the Wallenpaupack Power Co.

All this activity - and a shovelful of earth yet to be turned.

Ground was not broken for the power project for another 11 years. Then, in June 1924, construction men were on site in the Wallenpaupack River valley, beginning to clear the land. It took two years to build the dam and everything else necessary to generate 91 million kilowatt hours of electricity a year. The lake started to fill in 1926; it was stocked with fish provided by the Pennsylvania Fish Commission. Public enjoyment of the new lake, named Lake Wallenpaupack, began on August 1, 1928. *Today,* Lake Wallenpaupack, under the Pennsylvania Power & Light Co. (PP&L), has spawned a multi-million dollar tourist industry. At the same time, summer retreats are turning into year-round homes. The population of Wayne and Pike Counties is growing so rapidly that it makes northeastern Pennsylvania the fastest growing region in Pennsylvania. Additional private communities, built around lakes smaller than Wallenpaupack, have contributed to making Pike County the fastest growing county in Pennsylvania, and the second fastest growing county east of the Mississippi River. In addition to becoming the center for a large recreational community, Lake Wallenpaupack waters are sometimes used to control salt water advances in the lower Delaware River, in a cooperative program between PP&L and the Delaware River Basin Commission. As for the electricity generating capacity of "Lake Wallenpaupack," its original "raison d'être," that is now relegated to an emergency stand-by rôle.

Louis Watres died in 1937, having had an opportunity to see just how nicely Lake Wallenpaupack, the largest lake (man-made) within Pennsylvania's borders, had captured the imagination of vacationers. (Clarence Simpson died much earlier, in 1916.)

Again, this was the end of an era; this time, it was signaled by Watres' passing. For during World War II, beginning in 1939 and not ending until 1945, anthracite's worth became almost non-existent. The war years did bring back a temporary demand for hard coal; but afterwards, oil and what could be done with oil was far more desirable.

After World War II, there was no doubt about it, the United States was truly a world power and a world leader - The anthracite mines have fallen silent.

Partial Bibliography

Clark, J. A., ed. *The Wyoming Valley, Upper Waters of the Susquehanna, and the Lackawanna Coal-Region, Including Views of the Natural Scenery of Northern Pennsylvania, from the Indian Occupancy to the Year 1875.* Scranton: J. A. Clark, Publisher, 1875.

Crew, H. W. *History of Scranton, Penn.* Dayton: United Brethren Publishing House, 1891.

Goodrich, Phineas G. *History of Wayne County.* Honesdale: Haines & Beardsley, 1880.

Hollister, M.D., H. *History of the Lackawanna Valley.* 5th ed., Philadelphia: J. B. Lippincott Company, 1885.

Mathews, Alfred. *History of Wayne, Pike and Monroe Counties, Pennsylvania.* Philadelphia: R. T. Peck & Co., 1886.

Miner, Charles. *History of Wyoming, In A Series of Letters, From Chrales Miner To His Son William Penn Miner, Esq.* Philadelphia: J. Crissy, 1845.

Taber, Thomas Townsend. *The Delaware, Lackawanna & Western Railroad in the Nineteenth Century.* Muncy: Thomas T. Taber, III, 1977.

Taber, III, Thomas T. and Walter Casler and Benjamin F. G. Kline, Jr. *The Logging Railroad Era of Lumbering in Pennsylvania.* 13 vols. Muncy: Thomas T. Taber, III, 1970 to 1978.

About the Author

Aileen Sallom Freeman received her B.A. from the University of Pennsylvania and afterwards, studied at the Pennsylvania Academy of the Fine Arts. Her first foray into creative writing was at the age of eleven, when *The Grade Teacher* magazine published her play, *The Star Spangled Banner*. Now a professional artist, the author is married and has a son.